Open-Source Robotics and Process Control Cookbook

Open-Source Robotics and Process Control Cookbook

Designing and Building Robust, Dependable Real-Time Systems

by Lewin A.R.W. Edwards

AMSTERDAM • BOSTON • HEIDELBERG • LONDON
NEW YORK • OXFORD • PARIS • SAN DIEGO
SAN FRANCISCO • SINGAPORE • SYDNEY • TOKYO

Newnes is an imprint of Elsevier

Newnes

Newnes is an imprint of Elsevier
30 Corporate Drive, Suite 400, Burlington, MA 01803, USA
Linacre House, Jordan Hill, Oxford OX2 8DP, UK

 Recognizing the importance of preserving what has been written, Elsevier prints its books on acid-free paper whenever possible.

Library of Congress Cataloging-in-Publication Data

(Application submitted.)

British Library Cataloguing-in-Publication Data
A catalogue record for this book is available from the British Library.

ISBN: 0-7506-7778-3

For information on all Newnes publications,
visit our Web site at www.books.elsevier.com

04 05 06 07 08 09 10 9 8 7 6 5 4 3 2 1

Printed in the United States of America.

Dedication

This book is dedicated to my wife Cristen, in recognition of her uncomplaining acceptance of yards of PVC conduit in hallways, pounds of gel-cells in the living room, and never-ending snarls of wire and motors throughout the house.

Contents

Contents

About the Author

Lewin A.R.W. Edwards was born in Adelaide, Australia. He worked for five years in Melbourne, Australia on government-approved encryption, desktop protection and data security products for DOS, Windows and OS/2. For the next five years, he worked in Port Chester, New York for Digi-Frame, Inc., where he designed both the hardware and firmware of a range of multimedia digital picture frame appliances. These devices ranged in complexity from small pocket-size still-image viewers up to fully networked wall-mounted devices with audio and full-motion video support. He currently lives in New York City and works as a short-range radio digital design engineer for a well-known manufacturer of wireless security and fire safety products. His earlier works include *Embedded Systems Design on a Shoestring*, (a book about low-cost embedded systems development, principally targeted at ARM7 platforms), as well as articles on specialized design considerations for the microcontrollers used in electronic toys, commentary on Universal Plug'N'Play, reverse-engineering Internet appliances, and other topics of interest.

What's on the CD-ROM?

Included on the accompanying CD-ROM:

- A free version of the schematic capture and PCB CAD software used to prepare this book. (Refer to the license agreement included with the software for usage restrictions and limitations.)

- Atmel AVR Studio® 4.08.

- Full schematics and sourcecode for the projects described in this book.

- Ready-made disk images for the miniature Linux distribution used as a basis for the book's PC-side software.

- Distribution archives of the sourcecode for all GNU software used, along with application-specific patches, where appropriate.

CHAPTER 1

Introduction

1.1 History of this Book and What You'll Get From Reading It

Over the course of roughly a year, after completing my first book, I resurrected an old pet project of building an autonomous submarine (referred to as the E-2 project) with certain fairly challenging functionality requirements. In the course of developing this idea, I spent many hours on the Internet and elsewhere, researching techniques for rapid development of various electromechanical control systems and platforms to run fairly complex signal-processing algorithms. Although there are, of course, thousands of useful projects and snippets of information to be obtained from the Internet and books on hobbyist robotics, I found that nobody else seemed to have my exact priorities. In particular, there is apparently no single reference that gathers together at least introductory solutions to all the embedded design issues that affected my project: a need to use low-cost (open-source) tools and operating systems, a requirement for several features with fairly hard real-time requirements, and a desire to use cheap, off-the-shelf consumer grade components wherever possible. Available resources on many topics concentrate either on very expensive off-the-shelf industrial components, or on tightly constrained systems built around a single microcontroller, with delicately optimized, nonportable code to control peripherals—and a very limited range of peripheral support, at that. These latter system design restrictions are unavoidable when you're working to tight power requirements, space constraints, or a rock-bottom bill of material (BOM) cost, but it's an inordinate amount of effort to build and tune such systems for a one-off project or a prototype. Furthermore, learning all the details required to assemble such a system is an enormous task; it's easy to get lost in fine-tuning details without ever managing to field a complete, working system. Irritatingly, many of the tweaks and most of the careful planning you do to get that system operational will have to be thrown away if you move into actual production, or if you need to build some more units with slightly different components.

What I was searching for while developing the E-2 project was a way to build various hard real-time modules (sensors and actuators) that could easily and cheaply be interfaced to a general-purpose computer running Linux. The Linux box served as a testbed for algorithms which would later be ported down into a smaller, cooler, more power-efficient processing module of some kind. I needed a solid basis of known-good code and techniques so that I could strike out from that point and build my own customized system. I also wanted a simple up-and-running guide to building embedded Linux distributions. For the initial, nonfieldable prototype of my submarine, I didn't have an exact idea of how much CPU horsepower I would need in the final version—so I didn't want to get tied to a specific microcontroller architecture, nor did I want to get bogged down in trying to tweak and tune many real-time tasks on a single microcontroller. I also wanted to use a few peripherals—such as cameras —which are easiest interfaced through a general-purpose operating system.

These requirements may sound a chord with your own working life. Chances are you've encountered situations where it would be useful to automate some long-term data-gathering experiment or create a simple automated controller for a programming, manufacturing or other task. In this vein, three other instances where I have applied the techniques in this book are:

- The development of a range of successful (commercially-fielded) networked multimedia appliances, designed for unattended advertising and art-gallery applications.

- The development of specialized, almost wholly automatic mechanical failure testing apparatus for certain consumer electronics articles.

- Construction of an automatic high-speed datalogger that monitors a radio link and extracts DTMF messages targeted at specific receivers.

The second item above is of particular interest, because it demonstrates nicely how this book can be of practical value in process control and testing applications. During 2002, I briefly worked for a small division of a multinational company whose major focus was household and office plasticware. It was most instructive to examine their automated test fixtures—proprietary systems—and compare the cost and setup complexity of these fixtures with the relatively low cost and setup time of a simple Linux-based SBC controlling the same pneumatic actuators and sensors. Clearly, there is an under-exploited market for low-cost test systems of this type. The proprietary systems in use at this particular facility cost almost $20,000 for a half-dozen actuators and the associated PLCs, plus uncounted hours of setup time[1]. The control software for these devices was specialized and not well-understood; in fact, most of this equipment was standing idle because the people who originally configured it had left the company. By way of contrast, the same tasks could easily be accomplished with a regular PC costing a mere few hundred dollars, plus perhaps $200 per actuator for the associated pneumatics. More importantly, the control software for such a system is a simple C program easily understood and adaptable by any competent computer science or electronic engineering major; there were several candidates readily available in the company lab.

Due to the nature of the research which led to this book's inception, I have included a sprinkling of naval details within the text, not all of which are directly relevant to the embedded engineer. If this material is not of interest, you can safely ignore it without compromising your understanding of the remaining text in any way. The reason this information is included alongside the "pure" embedded development discussion is principally to illustrate the real-world requirements and thinking that led to various design decisions in the E-2 project. Engineering is not theoretical science; it is an applied discipline, and it is with this in mind that I use specific examples to illustrate theoretical points.

You should also note that some of the opinions expressed in this book, if not exactly controversial (except by Usenet standards—*everything* on Usenet is controversial!), are at least arguable; for example, the choice of AVR as my real-time control platform. For this reason, I have provided additional justification for the decisions I have made in this text. This additional explanation should demonstrate the reasons I had for choosing specific paths, but it's expressly *not* designed to prosely-

---

[1] The system was originally set up by "free" interns, so their time wasn't rigorously tracked.

tize the AVR to people who have experience with, and prefer, another architecture. Again, this "bonus material" is not critical to your understanding of the basic concepts presented here, and you can safely skip it if you wish.

Also keep in mind that this book is intentionally not a "bible." It is not an exhaustive coverage of every single nuance of the topics covered; such a work would span several shelves. The primary goal of this book is to describe and illustrate a simple, modular, inexpensive methodology for implementing complex embedded systems, and to present some ready-to-use modules for the reader to adapt to his or her own projects. The particular emphasis is on realizing end-to-end solutions using low-cost development hardware and free software tools. By the time you reach the last few pages, hopefully you should have the following:

- A functional understanding of the critical "under-the-hood" details required to bootstrap Linux on x86 platforms.

- An introduction to the types of problems you will face in using embedded x86 single-board computers as the core of data logging and motion-controlling systems.

- Basic information about the Atmel AVR microcontroller family.

- A practical introduction to building some simple data acquisition and motor control circuits, and connecting them to PCs.

- Some basic "primer" information on data security, authentication and reliability issues as they affect embedded systems.

The underlying idea is that the reader has reasonably intimate experience with one or other of the topics of Linux application development, or development of deeply embedded systems—this book is designed to boost you up the leading edge of your learning curve with the essentials of whichever side of the equation you're missing. It also provides the glue that binds these pieces of information together in the overall context of a fairly complex project. Note, by the way, that I used the titular word "cookbook" with some diffidence. Purely cookbook engineering—slotting ill-understood pieces together like Capsela spheres—is never good practice. In this book, I'm giving you some ready-to-use Capsela pieces, but I'm also telling you how and why I made the gears and shafts in each piece, and to some extent how you can go further and make better pieces for yourself. These explanations are much more important than the blueprints for the pieces themselves.

When planning a book like this, it's easy to fall into one of two traps: either to create a single, monolithic "mega-application" which illustrates all the desired points, but is extremely difficult to explain succinctly, or on the other hand to break the topic down into numerous small abstract notes that many readers will have trouble integrating into real-world projects. I have tried to steer between these two extremes by breaking the more interesting modules of the E-2 project into a few small, practical applications together with basically standalone code and enough theory to modify and extend these applications for your own uses.

Finally, a note to people who own my previous book, *Embedded System Development on a Shoestring*. This book is not designed to be a sequel to that volume, but it is definitely related material. If you follow the techniques in this book to build a prototype device, and you later want to squeeze it down into an optimized single-chip solution, my earlier work will help you understand how to use free GNU tools to get your software easily ported across to an ARM microcontroller. The principal criticisms I received for that previous book were that it needed to cover a wider range of information, and that there were too few illustrations, making it a rather dry read. I've listened to these comments and I hope you will find this book satisfies your needs in both respects. As always, your comments and suggestions are welcome; you can email me at sysadm@zws.com or visit my web site at *http://www.zws.com/*.

1.2 Target Readership and Required Skills and Tools

Throughout this text, I assume that the reader is a competent C programmer, with some experience in using (though not necessarily embedding) UNIX-like systems, specifically Linux. I also assume a very basic level of knowledge of real-time systems and simple digital electronics. This book is not an introduction to Linux, nor is it an introduction to the concepts of embedded programming; there are hundreds of such books already.

In order to follow along with the examples in this book, you will need the following:

- An x86-based PC system running Linux. This book was developed using Fedora Core 1, which can be downloaded for free (or purchased on CD) from *http://fedora.redhat.com/*. A full Linux distribution is not included with this book due to disk space constraints. For simplicity's sake, I suggest you use Fedora Core 1 unless you have special reasons for using a different distribution.

- Ideally, an x86-based SBC selected from the list in Section 2.5, with a hard drive and CD-ROM drive attached, and a CompactFlash® card of at least 8 MB capacity—however, none of these items are absolutely essential.

- A means for burning AVR microcontrollers. There are numerous schematics for simple AVR programmers available freely on the Internet, and a minimal programmer is simple to breadboard. (More on this in Section 2.3). I specifically recommend the STK500 development board from Atmel, because it is fully integrated with Atmel's AVR Studio IDE, and the $79 price is better value than the effort of building a comparable development system from schematics.

- An AVR development environment, or at least an assembler. The projects in this book were developed using the free Windows®-based AVR Studio® from Atmel, which is included on the CD-ROM. Pure-Linux shops may prefer to use the free avrasm assembler, which I have also included. The avrdude package can be used to burn chips under Linux.

- An oscilloscope is highly recommended, though not mandatory. When you're debugging serial communications protocols in particular, nothing beats being able to see the actual bits going to and fro. The waveform screenshots in this book were taken using a Tektronix TDS210 60 MHz two-channel digital scope.

1.3 Conventions Used in the Text

Throughout this book, I have attempted to adhere to a single set of conventions:

- For the sake of consistency, all measurements are given in metric units, *except* where they refer to a controlling dimension that was originally specified in nonmetric units. In the latter case, the controlling dimension is specified in its native unit, and the metric equivalent is shown in parentheses, as in "The length of a foot-rule is 12 inches (30 cm)." In some cases, this results in a departure from accepted industry standards; for example, the speed of a seagoing vessel is normally specified in knots (nautical miles per hour), not ms^{-1}.

- URLs are written in italics, for example, *http://www.zws.com/* in order to separate them from surrounding punctuation.

- In common with most other technical publications, sourcecode and command line text that is intended to be typed verbatim is rendered in a `fixed-space` font.

- Occasionally, you will find UNIX commands and library functions mentioned with the standard nomenclature of command(n), where n is the section containing the manual page for the command in question. For example, you can find out more about rdev(8) by typing `man 8 rdev` at a shell prompt. Don't go looking for nonexistent footnotes!

- When I discuss procedures that access the enclosed CD-ROM from Linux, I assume that the disk is mounted at /mnt/cdrom, because that is where most desktop Linux distributions will put it. If you mount it somewhere else, you'll need to edit your command-line entry appropriately.

- All sourcecode and makefiles are designed to be viewed with a tab width of four character spaces.

CHAPTER 2

Microcontrollers, Single-Board Computers and Development Tools

2.1 The Division of Labor

The designer of a complex multi-level project such as the E-2 must frequently juggle the following conflicting requirements, among others:

- Hard real-time response requirements of sections of the overall system.

- The hardware and firmware complexity of interfacing special peripherals such as cameras, Ethernet networking, 802.11b wireless networking, and others.

- Bill-of-materials costs for both prototypes and production pieces.

- Development time.

- Cost of development tools.

- Relatively high cost of components designed for embedded systems, as compared to the pricing of comparable-performance, generally-available consumer products.

It's a terribly daunting task to approach all of these problems at once, particularly at the start of a project when your exact needs are generally not well-specified. Limited time or monetary budgets add stress, because there simply may be no days or dollars spare to be wasted exploring dead-end research paths on the way to a working system. Furthermore, many of the systems of interest to readers of this book will either be unique, or will be produced in very small volumes. For such systems, it's hard to justify intense time expenditures researching and fine-tuning noncore features (i.e., the infrastructure features you have to debug before you can debug the functionality you actually want to develop).

The basic methodology I have used to cut through most of this Gordian knot is as follows: To begin with, I divide all the system processes into two categories, which I will term "hard" and "soft." For the purpose of this discussion, hard processes are defined as direct physical-world interaction tasks where timing and system robustness are likely to be critical to performance and/or safety. Examples in the E-2 system are: Stepper motor control for rudder and dive planes, battery charge and thermal monitoring, depth monitoring, propulsion motor control, and bilge sensors. Hard processes are typically easy to identify and characterize precisely, and can often be implemented in a small 8-bit microcontroller. In the E-2, we will perform all the hard tasks using microcontrollers in the AVR series, from Atmel.

By contrast with hard processes, soft processes are not at all mission-critical, and have relaxed or nonexistent real-time requirements. Generally, soft tasks can crash, provide erroneous, untimely or downright missing data, and the overall system health will not be unduly compromised. Examples in the E-2 project are image capture, storage and analysis, data logging, and some telemetry functions. Many soft tasks require interaction with complex sensory or communications modules such as cameras and wireless networks. For this reason, it is convenient to use standard off-the-shelf consumer peripherals such as USB webcams, CompactFlash storage media, USB wireless LAN pods, and others. Interfacing to these sorts of peripherals from a small microcontroller is often decidedly nontrivial—oftentimes, technical data is hard to come by, and it's also frequently difficult to acquire loose sample parts in small quantities. Prototyping with these parts is also usually difficult.

In the case of low-cost CMOS image sensors, for example, virtually the only way to get these parts off the shelf is to buy a complete camera and cannibalize it—IF you can identify the devices in it without microscopic examination, and IF you can get datasheets! Furthermore, manufacturers of consumer electronics are more or less constantly refining and costing down their products. You may cannibalize MyWidget V1.0 and spend many hours getting the components to work in your system, only to find (when you start to build a second unit, e.g., to replace a lost prototype) that MyWidget V1.0 has been superseded by V1.1, containing totally different components—maybe even an undocumented ASIC.

In a similar manner to the way I handled task management, I divide system communications into two classes; control-critical and noncritical. In the case of E-2, all control-critical data transfers occur within the vehicle itself, between the various real-time modules and the main controller. These communications take place over an internal three-wire SPI-style serial bus. Noncritical communications are, for example, the ship-to-shore telemetry link. These data streams can be carried using whatever media and protocols are convenient, with less attention paid to real-time issues such as lag[2].

You may wish to pause here and consider the implications of the preceding decisions. In particular, note the implication that hard tasks and control-critical links are trusted and soft tasks and noncritical links are not trusted. We're going to be running all the hard, critical stuff in small microcontrollers carefully programmed "to the metal," and—hopefully—completely understood and characterized in all conceivable situations. The messy stuff like networking, snapping pictures to a hard drive, and so on, will all be run on a totally separate piece of hardware. If it crashes, it can simply be reset or shut down with no impact on system survivability. This is important, because most of the software running on that untrusted piece of hardware wasn't developed for embedded use, and it certainly isn't as well-defined as the software we custom-engineered into the hard-task controllers.

I should stress that none of the previous discussion is *per se* an indictment of the reliability of embedded Linux. It is perfectly possible to build rock-solid control systems based around a single Linux processor, and there are many such systems in existence. However, a uniprocessor system requires considerable fine-tuning of the operating system and application software to achieve a sufficiently real-time end result[3]. Furthermore, in order to achieve such a result, it is often necessary to use nonstandard software components intended specifically for embedded systems (real-

---

[2] Obviously, this isn't true of all telemetry applications. In E-2, the telemetry signal is provided solely as a convenience to the shoreside operator; it's not critical that it be strictly real time or that it implement strenuous error correction.

[3] Uncharitable people say of embedded Linux that the standard development technique is to write the device driver or application the way you think it should be written, then add hardware until it performs successfully, to the desired approximation of "real time." The fact that this is so often true is more an indictment of the developers than the OS, though.

time Linux extensions such as RTLinux, for example). The net effect of both of these factors is greatly to increase development time, and generally also to tie you to a specific hardware/software combination. The major advantage gained by the dual-tier, trusted-vs.-untrusted layer solution is the ability to lash together functional, but hard-to-guarantee features on the untrusted layer, using off-the-shelf software and hardware components.

The crucially important technical advantages of our method of putting together our complex embedded system are, therefore:

- The real-time characteristics of any given hard module can be tuned right down to the CPU-cycle level, if desired.

- Changes to any one real-time module don't directly impact the timing properties of any of the other modules.

- Standardizing communication protocols amongst the various modules establishes a "firewall" of sorts, which is useful both for testing purposes (as-yet-unbuilt modules can be simulated with a piece of external hardware) and for future upgrades (modules can be replaced with updated versions as long as a consistent software interface is maintained).

- Reuse of hardware modules in other projects is very easy, since it is the *system* that is project-specific, not the individual parts.

- Because access to complex peripherals is abstracted at a fairly high layer (through the operating system running on the untrusted soft-task controller), it's possible to swap out these components for functionally equivalent parts without writing custom device drivers.

In fairness, at this time I should also point out the downsides of the multi-module way of doing things:

- The overall bill-of-materials cost for a multi-module system is likely to be much higher in the long term. This is not likely to be a big factor for proto-type or short-run construction, where setup costs dominate the unit price. For mass-production, however, the price advantages of a uniprocessor system become progressively more attractive. Note, though, that for low-volume or unique applications, the higher BOM cost of the multiprocessor system may be partially or completely offset by the high cost of obtaining required evaluation hardware for the devices used in the "cheap" uniprocessor design, so the development method I describe here is likely to work out cheaper, per unit, for low-volume designs.

- Power consumption and physical size will be larger than for a fine-tuned system.

In closing this section, I'd like to rebut one commonly-raised argument against multiprocessor systems: Many people believe that by introducing multiple microcontrollers, you are increasing the number of possible points of failure and thereby making the overall system inherently less reliable. The most succinct counter-argument to this, to which I subscribe, is that conceptually, the same bugs and design shortcomings will exist whether a particular set of features A, B, C are implemented on one processor or three individual processors. Keeping the three functions physically separate prevents them from interfering in each others' address spaces, and also allows fast system recovery—because if, say, processor B crashes, processors A and C can continue to operate unaffected while B reboots.

It is, of course, true that adding more silicon increases the possibility of "SEUs" (single-event upsets) caused by environmental stresses such as incident radiation, simply because there is more silicon real estate to be affected by such factors. This is, however, a relatively subtle point and is unlikely to be an overriding concern in the majority of systems to be built by readers of this book.

2.2 Candidate Microcontrollers for 'Hard' Tasks

Given that we need to choose a microcontroller family to handle the real-time parts of our system, let's first create a short list of rules for selecting this family:

- Assemblers and compilers must be freely available, either from the manufacturer or as a result of open-source efforts such as gcc.

- Device programming hardware must either be low-cost or simple enough to build at home using off-the-shelf parts.

- Parts to be used must be available ex stock from major mail-order distributors such as Digi-Key, Newark, and others, with no minimum purchase requirements.

- Device family must contain parts spanning the widest possible variety of ROM, RAM and peripheral requirements, with as much firmware and hardware design commonality as possible.

- Ideally, the parts chosen should enable easy implementation of a slave SPI interface, but this isn't vital (and SPI is extremely simple to bit-bang, anyway).

There are three obvious targets that present themselves immediately: 8051, Microchip PIC®, and Atmel AVR®. The ancient 8051 is indubitably the world's best-known candidate for 8-bit applications, so we'll start by examining this family briefly. It's very inexpensive, available from an unparalleled number of sources (Atmel, Philips, Winbond, Cypress, and Dallas/Maxim are just a few of the vendors with standard 8051 parts; dozens more have 8051-cored ASICs and ASSPs), and the basic architecture is familiar to most embedded engineers. There are numerous high-quality tools and reference designs, and megabytes of sample sourcecode available.

The main reason I have chosen to avoid the 8051 family is because of the lack of standardization across manufacturers. No single manufacturer carries an 8051 variant to suit every single application need, and almost every manufacturer has added somewhat proprietary features to the core or peripherals. Because of the long history of this part, it is even common for a given manufacturer to have two or more completely different lines of 8051-cored parts, with different family trees, idiosyncrasies and programming hardware and software tools. Some 8051 sub-families require external programming hardware; some have in-system programming capabilities, many do not have flash memory, and in order to migrate from one variant to another may require investment in relatively expensive programming hardware. It's possible to avoid some of this nonstandardization by sticking to a set of "vanilla" 8051-cored parts that are implemented nearly identically across manufacturers, but this also means avoiding use of most of the 8051s with interesting nonstandard peripherals; LCD controllers, USB, on-chip A/D and D/A converters, expanded ROM or RAM, in-circuit programming, etc. It also means that, in a modular design where each microcontroller has minimal duties, you will likely be spending far too much on over-specified microcontrollers. For instance, you don't need kilobytes of RAM or ROM for a simple stepper motor controller!

As a secondary, but still relevant point, the 8051's architecture is positively archaic. The upside of this is that compiler vendors understand it very well, and commercial compilers for the 8051 are about as good as they're going to get. The downside is that even the best 8051 compiler (arguably, Keil's product) is unavoidably less efficient than good compilers targeted at more modern processors. Worse still, the only halfway decent open-source C compiler for the 8051 (sdcc) is exactly that—only *halfway* decent. And writing and maintaining large volumes of 8051 assembly language is irritating. It's an entirely justifiable effort if you're making large volumes of something or have another good reason to pick that architecture, but if you're trying to follow the path of least resistance to build a low-volume system with the minimum possible personnel resources, other microcontrollers are a better investment.

In my opinion, therefore, 8051 variants are a great choice when you have a specific application in mind, and you are looking for a one-chip solution. Because of the anarchic differences between different vendors' sub-families, and the fact that no single vendor carries completely code-compatible parts to suit every application, I feel that 8051 isn't such a good choice for modular applications where you anticipate the need to use many tiny microcontrollers in a single project. The workload required to keep code mobile amongst different 8051 variants with disparate peripherals is quite significant. If, however, you are experienced with the 8051, there is no reason why you can't apply that knowledge to the techniques in this book.

For the projects you will find here, I have chosen to use the Atmel AVR series of microcontrollers. These parts are all flash-based; the family offers a reasonably wide range of functionality, and the instruction set is easy-to-learn and to a large degree common amongst family members. Under most circumstances, AVRs are programmable in-system or in an external socket using a simple-to-manufacture parallel port cable. The official STK500 development board, should you wish to acquire it, is cheap ($79 is the current list price) and fully-featured. A functional Windows IDE and assembler are free from Atmel, a port of gcc is also available and supported by Atmel, and there are freeware assemblers and other tools for UNIX-based operating systems as well as Windows.

Another ubiquitous microcontroller family, commonly used in low-volume and hobbyist applications, is the Microchip PIC. This family meets essentially all of the requirements in the preceding list. I have not chosen to use it, however, simply because it is slightly harder to learn and use than AVR. (By the way, I base that comment on my own experience in learning the two cores, as well as commentary I have read from neophytes asking for help and advice. This is, however, one of those potentially controversial topics I warned about in the introduction. I'm certainly not condemning the PIC as a hard-to-use maverick, I'm simply pointing out that many people seem to find the AVR family easier to use). One other downside to the PIC family is that the "official" entry-level development kit (PICstart Plus) is more expensive than the STK500—almost three times the price, in fact—and it's nowhere near as flexible, being simply a dumb chip-burner with no prototyping functionality at all.

There are a couple of other reasonably popular microcontroller families that we could have considered, and you may wish to investigate them yourself. The Texas Instruments MSP430 family, for example, is a very interesting range of parts. It combines a 16-bit RISC core (some variants have a bonus hardware multiplier) with various useful peripherals, at an attractive price point. The parts are flash-based and support JTAG debugging using an inexpensive parallel-port or USB-based wiggler; a most useful feature. The downsides to the MSP430 are prototyping issues due to the small packages used, and also interfacing problems arise due to the fact that they are 3.3V parts. However, if you're trying to cut down your power budget, or you're looking for a high-performance core that's inexpensive and well-supported by a major vendor, MSP430 is a good choice.

Another micro that is worth at least a quick look is the range of 8-bit devices from Rabbit Semiconductor, *http://rabbitsemiconductor.com/*. These parts are derived from the ZiLOG Z-180, so depending on your background you might not have too much of a learning curve. They are firmly targeted at connected applications; Rabbit supplies a free TCP/IP stack and provides several evaluation boards and fairly low-cost, end-application-integratable CPU modules, some of which have Ethernet onboard. They even have a Wi-Fi kit, although it's rather expensive. The main downsides to Rabbit are the small size of the company, which argues against long-term availability (however, they have been around for several years and seem to enjoy good popularity in the hobbyist market), and the fact that their free "Dynamic

C" compiler is horribly nonstandard; it's tedious and most inelegant to port code into or out of a Rabbit design. There is an ANSI C compiler available, but it is buyware. Arguments in favor of Rabbit are low entry cost (all the basic tools are free and the development hardware is reasonably priced), ease of low-volume manufacture (since Rabbit supplies the chips ready-to-run, already soldered down to a board, if you wish), and a rich feature set (large flash memory, large RAM, fairly simple programming with a C-like language as well as assembly language, and a lot of ready-to-use application-specific code, particularly in the realm of TCP/IP networking protocols). Possibly the most compelling argument for Rabbit, however, is the fact that you can migrate from one-time prototype production directly to low-volume manufacturing (a few hundred pieces a year, perhaps) without any need to redesign.

2.3 The Atmel AVR and its Development Hardware Up Close

After some careful thought about the pros and cons, I have decided to use a single type of AVR chip for all the example projects in this book but one. The reasons for this are twofold: first, to reduce the number of separate parts you need to acquire in order to build these projects (and to allow you to use the same chip for different projects, if you wish), and second to avoid too much explanatory text devoted to pedestrian compatibility issues. The particular AVR I have chosen is the ATtiny26L, which provides a good cross-section of the peripherals available in the AVR family. Migrating code snippets to other AVRs is not difficult.

The AVR series consists of a fairly broad range of hybrid-bit-width microcontrollers (nominally 16-bit code word, 8-bit data bus and ALU) sharing a common instruction set and differing primarily in the on-chip peripherals and package options. These devices don't show a clear genealogical relationship to any other microcontroller core I'm aware of, but some variants do show superficial signs of having been designed for people migrating away from the 8051 (the 40-pin AVRs are in a very similar pinout to a standard 40-pin 8051, for instance). AVR is a Harvard-architecture RISC core with 32 8-bit general-purpose registers, named R0–R31. These registers are mapped into the core's data address space at address $00-$1F. Registers R26–R31 have a secondary function for indirect addressing modes; they are divided into pairs named X (R26–R27), Y (R28–R29) and Z (R30–R31). Any of these three paired registers can be used as a 16-bit pointer into data RAM (the first register named is, in each case, the less significant byte of the address word). Most instructions can operate on any register; a few instructions (such as word-add, word-subtract, and load immediate) can operate only on a subset of the registers, R16–R31.

The AVR core also has a separate 64-byte I/O address space to interface with the on-chip peripherals. All of these peripheral control registers are conveniently mirrored in the general data address space at locations $20-$5F, so that you can access them with different addressing modes if you wish. The ATtiny26L also has 128 bytes of SRAM from $60-$DF, and the remainder of the data address space is unimplemented. Unlike PIC variants that have a limited-depth hardware stack separate from the processor's other address spaces, AVR supports a traditional stack in the on-chip SRAM. The stack pointer is simply an 8-bit register in the I/O address space.

Some other features of the tiny26L, in no particular order, are:

- 128 bytes of EEPROM, useful for storing configuration and calibration data, or failure information for postmortem analysis.

- A simple but very flexible "USI" (Universal Serial Interface) peripheral, configurable to act as an I²C, SPI or asynchronous serial port. For trademark reasons, the I²C mode is referred to in Atmel literature as Two-wire, and the SPI mode is referred to as Three-wire.

- Two timer/counters, configurable in a variety of modes. One of these timers can be programmed to provide two PWM channels with positive and inverted outputs.

- Eight analog-to-digital converter channels.

- Brownout detector, configurable for 3.3 V or 5 V operation, and watchdog timer.

- In-system programming capability using the built-in SPI interface.

One important fact to note about in-system serial programming is that the microcontroller needs to have a core clock source. Simply providing the SPI data clock is not enough! This means that if you're tinkering with the fuse settings, you have to be careful that you don't disable the system clock. The designs in this book all use an external crystal oscillator. It is unlikely, though not entirely impossible, that you'll get yourself into a "clockless" situation with such circuits. However, in designs that use the AVR's internal RC oscillator and that re-use the clock input lines for other functions, there is a real hazard that you can disable the device by selecting an external clock mode. To recover from this, you can tristate your external hardware (or lift a CPU pin) and feed in an external clock temporarily.

In the same vein, AVR fuse settings allow you to disable the reset input and use the pin as a GPIO. If you do this, you cannot use serial in-system programming; you must use a parallel programmer. The STK500 is a suitable piece of hardware.

These issues are by no means unique to the AVR family; most microcontrollers that support in-system programming have the same sort of limitation. These problems are also not, as a rule, very important for hobbyist circuits, which typically use socketed DIP microcontrollers. Once you start etching PCBs for your designs, however, it becomes very attractive to use surface-mount packages for size and cost reasons; be careful not to paint yourself into a corner when you're upgrading the firmware on an assembled PCB. By the way, note that Atmel ships AVRs from the factory with an internal RC clock source selected by default, so that you can stuff your board with blank, factory-fresh chips and program them later over the SPI interface.

What about firmware development tools? There are a number of products you can use for compiling and burning AVR code. In order to make the example source-code in this book as easily portable between toolchains as possible, I have written it entirely in assembly language. The software development environment I used was Atmel's free AVR Studio for Windows, version 4.08, in conjunction with the STK500 evaluation board. AVR Studio is included on the CD-ROM with this book in the "utils/AVR Studio 4.08" directory, and I strongly advocate using it. However, if you need to use a different assembler (for example, if you're developing under Linux), please try to use the standard Atmel include files, or at least duplicate whatever snippets you need, rather than writing your own set of symbols to describe the registers in the chip. It will be very annoying—to *you* as well as to other luckless souls updating your work—to have to port code to another member of the AVR family if you use hand-rolled register and bitfield names.

The STK500 is a very flexible, serial-controlled development board that directly supports all of the DIP-packaged AVR chips and, with the STK501 adapter board, the larger 64-pin surface-mount parts. It can be used to burn microcontrollers inserted in the DIP sockets on the STK500 itself, or it can burn devices already mounted onto your subassemblies, via an in-system-programming cable. It sports eight push-buttons and eight LEDs, and it brings all the I/O lines to 100 mil headers - so you can do some or all of your code debugging directly on the development board. The onboard supervisor microcontroller that manages the STK also allows you to program various clock rates for the device under test, which is a boon to debugging some types of problems—bringing the clock down REALLY slow lets you examine signal state changes in slow motion. For those of you struggling under the evil oppression of a legacy-free PC with only USB ports, the STK500 also works perfectly over a USB-to-serial adapter.

One aspect of the STK500 that is slightly unusual is that you have to set up—by hand—the connections between the device under test and the clock/programming nets required to access it. This is not documented as well as you'd probably like—and it isn't documented at all for some new devices like the ATtiny26L, at least at the time of writing. For the projects in this book, you should take note of the following:

- Your ATtiny26L chip should be inserted in the blue socket labeled SCKT3700A1.

- The ISP6PIN header should be jumped to the SPROG1 (blue) 6-pin header.

- XT1/XT2 on the PORTE/AUX header should be jumped to PB4/PB5 (respectively) on the PORTB header.

- RST on the PORTE/AUX header should be jumped to PB7 on the PORTB header.

- While I was testing the code in this book, I generally had PORTA jumped to the LEDS connector, so that LED0-7 reflect the state of PA0-7, and I jumped SW0-3 on the SWITCHES header to PB0-3 on the PORTB connector.

- Jumpers should be set as follows: VTARGET, AREF, RESET, XTAL1 all shorted, OSCSEL pins 1-2 shorted, BSEL2 open.

All the cables required to perform the above interconnections are shipped as part of the STK500 package.

A final note on AVR Studio: The current version of this program can be rather sensitive to the presence of software that installs filesystem hooks. If you are having difficulty building code (typically, the symptom you will get is that you hit F7 to build, and nothing appears in the output window), try disabling any antivirus software or automatic file versioning utility you have running in the background. This conflict is known to occur, on some systems at least, with both Norton Antivirus and Vet.

2.4 Candidate x86-based SBCs for 'Soft' Tasks

The reason for choosing an Intel-type PC-compatible SBC rather than a proprietary or semi-proprietary architecture based around some RISC microcontroller is primarily ease of development, followed closely by cost. There are numerous readily-available RISC-cored system-on-chip devices (and SBCs based on these parts) which would have adequate processing performance for the E-2 project, and MUCH leaner power requirements. However, the SBCs based on these devices are, by and large, very low-volume, expensive appliances, and developing for and interfacing to them presents significantly greater engineering challenges than simply attaching peripherals to standard PC ports and installing a pre-built driver. By using a hardware platform that is essentially just an off-the-shelf PC-compatible, we can concentrate on the application at hand, rather than spending time on creating a toolchain, configuring and compiling a compatible kernel, and working out the minutiae of interfacing the peripherals we need to use. Our development process is thereby greatly accelerated; refer to the next chapter for a more detailed analysis of this point.

If your requirements are such that you absolutely MUST have low power consumption in the master controller, then you do have a few options. Several companies—including Advantech—sell SBCs based on Intel's XScale® CPU; for example, look at the VIPER product from Arcom, *http://www.arcom.com/*. These boards are generally built on the standard 3.5" biscuit form factor (see the following) and are supported with ARM-Linux. If you are willing to consider more deeply-embedded solutions running leaner operating systems, there are even more options for you, such as the LPC-xxxx series of evaluation/prototyping boards from Olimex, *http://www.olimex.com/*. These boards are based around the new Philips LPC2000 series of ARM7-cored microcontrollers; they're supported by GNU tools, simple to program, and the offerings from

Olimex are very reasonably priced (around $60). Due to RAM limitations, you won't be able to use Linux on these boards; they're best suited to proprietary OS-less environments, or very small operating systems such as uCos-II.

Our selection of x86 leaves us a lot of territory from which to choose, however. There are numerous vendors offering single-board x86-Linux-compatible computers based around processors ranging from the 80386 (in the form of the Intel i386EX embedded controller) all the way up to high-end multiprocessor Pentium 4 and even 64-bit boards[4]. These boards are readily available in a variety of largely standardized form factors:

- **3.5″ biscuit.** This form factor has the same footprint as a 3.5″ disk drive. Power input is via a 4-pin connector carrying +5 V, +12 V and two ground returns, the same type found on a hard disk or CD-ROM drive.

- **5.25″ biscuit.** This form factor has the same footprint as a 5.25″ disk drive. Generally, the power input is via an old AT-style (not ATX!) connector.

- **ISA or PCI processor module cards**, intended to plug into a passive backplane alongside peripheral cards with the same bus architecture. By the way, a common misconception is that multiple CPU cards of this type can be plugged into a single backplane to build a multiprocessor system; this is never the case for ISA boards, and only occasionally true for PCI cards. Unless the card's documentation *specifically* says that it's designed for use in a multiprocessor environment, you should assume that it can't operate this way. Even if it is possible to build multiprocessor systems around a particular CPU card, a specialized backplane will almost certainly be required. In many cases, these "multiprocessor" backplanes actually have no common connections except the power rails; any inter-processor communications you wish to implement have to be routed through Ethernet or some other user-supplied interconnect mechanism.

- **Mini-ITX motherboards.** This form factor is mechanically a subset of the standard ITX board used in desktop PCs, and it has a connector for a standard

---

[4] Some vendors still sell systems based around older x86 processors—80186-compatibles are quite common—but we won't discuss these.

ATX power supply. Mini-ITX implementations extant at the time of writing require +3.3 V, +5 V and +12 V rails.

- **Standard-sized PC motherboards** with varying levels of on-board peripheral integration.

The Mini-ITX form factor mentioned previously straddles the line between the "consumer off-the-shelf" and "embedded" markets, and deserves a little additional explanation. At the time of writing, the major vendor of Mini-ITX boards is Via Technologies, *http://www.viavpsd.com/*, but other manufacturers are preparing to release similar products. Among these is Transmeta, who have chosen the Mini-ITX form factor for the evaluation boards for their newest x86-compatible processors. Mini-ITX is a physically cut-down (170 × 170mm), backwards-compatible version of the ITX motherboard form factor; it has screw holes and connector zones designed to mate with a standard PC casing and ATX power supply. Via Technologies vigorously markets this form factor to a sector one might characterize as "consumer embedded" applications; i.e., hobbyist projects built around a PC-compatible motherboard. The TV-out feature included on Via's Epia Mini-ITX range has led to a large number of hobbyists using these boards to build dedicated set-top boxes for playing video content downloaded from the Internet. There are also quite a few commercial thin client sorts of applications built around these boards.

Via's latest mini-ITX boards are much more embedded-friendly than the older boards (which were basically just a regular PC motherboard writ small). The latest models have PCMCIA and CompactFlash slots and an even smaller outline than Mini-ITX (Via terms this "Nano-ITX"); they are also substantially cheaper than standard 3.5″ and 5.25″ SBCs based around the exact same chipsets. Speaking of prices, just as a data point for you, Mini-ITX boards start at just under $90 retail, single-unit pricing, for a complete board with 533 MHz Via Eden CPU and various integrated peripherals; 3D accelerated AGP Super-VGA, two IDE buses, serial, parallel, Ethernet, four USB ports, etc—just add RAM. Pentium-class SBCs (of comparable performance) in 3.5″ or 5.25″ form factors start at just above $350 with a similar set of peripherals. However, that isn't the whole story. One major downside to Mini-ITX is that it assumes the availability of an ATX-style power supply. Via's boards, for instance, absolutely require +3.3 V, +5 V and +12 V rails—they won't operate without all of these voltages present. Most SBCs are happy with a single +5 V rail[5]; they have onboard regulation to provide whatever other core and I/O voltages they require. You should take the cost of a suitable power supply into account when building your system—and also consider issues like size, airflow/airspace requirements, and noise from cooling fans. Remember that most active cooling systems work constantly to pull any dangerous aerosols or dust in the atmosphere right through your system! My suggestion, if you plan to build Mini-ITX into your system (assuming you don't want to design your own power supply from scratch) is to look at the power supply modules manufactured for 1U rack-mount cases. These are standardized in size (hence, interchangeable) and most of them have variable-speed fans, which run only when the power supply is actually in need of active cooling. They also have enough power capacity to supply any peripheral you are likely to add to such a system.

One final comment on Mini-ITX and Nano-ITX—The star of this form factor is most definitely rising. Several manufacturers produce a range of standardized housings, power supplies and slim peripherals designed specifically for Mini-ITX boards. Some of these are intended for "thin client" diskless applications, others for semi-industrial rack-mount installations, and some of them even for set-top-box use. If you want to use the minimum possible quota of custom parts in your design, Mini-ITX is a great path to investigate.

A very important factor which you should always keep in mind is that consumer-grade PC products are constantly changing. It's extremely difficult to standardize a product if you're building it out of ill-specified parts, and that translates into ongoing costs for you in revising housing designs, re-testing your external circuit and firmware with different motherboard chipsets, and so on. Obviously, this isn't much of an issue for a one-off project, but it is a major sourcing issue for low-volume, long-term ongoing production, where your order quantities aren't high enough to guarantee supply of older parts. As a rule of thumb, if you anticipate the production life of your device

---

[5] Most (if not all) SBCs have inputs for at least +5 V and +12 V; many have inputs for negative rails too. However, in most cases you'll find that these extra voltages are only passed through to expansion ports; they're not actually used on the SBC. The PCM-5820, for instance, relies only on the +5 V rail. If available, it can use the +12 V rail to achieve a wider swing on the audio outputs, but if you don't want to provide a dual-voltage power supply, just set the audio jumper for "no 12 V" and the board is quite happy to run off +5 V only.

to span six months or more, and it has to be squeezed into a special form factor of any sort, then I strongly advise you to design around an SBC rather than an off-the-shelf motherboard. You'll pay something of a premium, but it will save a lot of time and money in the medium to long term because you won't need to revise cable harnesses, housing designs, power supply requirements, and so on. Remember also that in most jurisdictions, you have to pay for EMI compliance testing every time you make a design change like this, or risk enormous fines!

If you're willing to shoulder the annoyance of (at least potentially) keeping track of several software versions, and your application is not tightly space-constrained, then standard PC motherboard form factors do have a certain appeal—the ATX standard (and by extension, Mini-ITX) specifies a single standardized square cutout connector zone. Every motherboard ships with a small springy steel plate that mates with this connector zone and provides precise cutouts to match whatever connectors are provided on that specific motherboard. This neatly takes care of the EMI problems I mentioned earlier—the biggest annoyance (besides the fact that ATX motherboards are relatively large, and the volume of airspace you need to keep clear to match the ATX clearances standard fully is pretty vast) is that as supplies of a particular motherboard dry up, you'll need to test and qualify your software distribution on new platforms. If your application is the sort of animal that requires ongoing software updates, you have to decide whether to make a super-intelligent software upgrade bundle that can work out what kind of hardware it's running on and configure itself appropriately, or keep track of which users have which hardware versions. The latter approach is easy while you have only a few customers, but once your userbase swells, it becomes a big exercise in database management, *particularly* once a few units have been in for repair and have had parts "upgraded"—because you can no longer use simple serial number range checks to know what's inside a particular unit. Your situation may have special circumstances, but when I've been involved in a project like this, it has always been less work, ultimately, to build a single software bundle that works on all supported hardware versions. At the very least, design your system in such a way that it's possible for the software to determine what kind of hardware it's running on *before* it needs to do any hardware-specific startup tasks. That might sound crazy, but unless you design with this idea in mind, it's not uncommon to run into chicken-and-egg situations where the only way you can identify some piece of hardware is by assuming the existence of some other piece of hardware, probing which may crash the system if it doesn't contain this particular device.

On the topic of housing your system, if you're building a device in any reasonable production quantity, you will want a professional-looking enclosure for it. Jiffy boxes work, but they're ugly. Unfortunately, the tooling cost for a custom plastic enclosure is prohibitive—tens of thousands of dollars at minimum. A much cheaper option, which people rarely seem to consider, is bent sheet-steel. Numerous metal enclosure shops can build you quite complex shapes at a surprisingly low cost. Modern metal-working shops use CNC laser cutting tools on the raw sheet stock to make holes and tabs of practically any shape. The parts are then bent by hand, and spotwelded where necessary. Fasteners—threaded posts ready to accept a nut, tapped posts to accept a screw, reinforced spots for rivets, and so on—are permanently bonded to the sheet parts using a press apparatus. The parts can then be painted and baked, if desired. As a data point for you, a production run of around one hundred pieces, with an exterior paint job and approximately the same complexity as a desktop PC's casing, manu-factured locally in the United States, will cost around US$60 per piece. It's entirely possible to get even cheaper prices if you shop around—there are literally thousands of metal shops that can do this sort of work.

Another possible housing solution is to use a section of aluminum extrusion with custom-punched end plates. This type of housing works very well with 3.5″ SBCs and other boards that run all their important connectors to an edge. It's less work-able when you need to mount lots of connectors on the end panels using flying leads. (Cheap external modems were often manufactured using this method a few years ago). If you're contemplating this option, you may want to visit *http://www.frontpan-elexpress.com/*, where you can download software to create your custom end-panels and get an instant online quote on production.

In any case, before you sign off on your custom enclosure solution, compare your design with standard products and decide if the customization you've added is re-ally worth it. Remember—if you're using a Mini-ITX board, there are numerous low-profile housings available to you off the shelf in a variety of shapes. You can also think about using a standard 1U 19″ ATX rack-mount casing, which will already have a standard connector zone cutout in the back and a suitable, UL-listed power supply—plus, it has the advantage of integrating neatly with a lot of other industrial equipment. Another subtle advantage of this approach is that if you build a computer system out of FCC-approved parts, you don't need to seek a separate approval for the assembled system—it just rides the approval of its individual components.

2.5 The Advantech PCM-5820 Single-Board Computer Up Close

For this text, I have chosen to use the Advantech PCM-5820 single-board computer as my reference platform. This board has a combination of factors working in its favor, which is why I have been working with it for a couple of years:

- It is readily available ex-stock at a reasonable price (around US$235 at the time of writing; cheaper than many industrial SBCs of much lower performance).

- It is physically quite small, at 145×102 mm (it is a 3.5″ biscuit) and it is easy to mount.

- Its power requirements are relatively modest for a Pentium-class x86 system; Advantech quotes typical current requirements of 1.5A from a single +5 V rail, although peak current requirements can be as much as 4A. A side benefit of this is that it does not absolutely require active cooling (Advantech ships it with a thin passive heatsink); as long as you don't actually wrap it in a blanket, overheating is unlikely to be a problem.

- The board sports a healthy selection of peripherals and I/O features, making it very easy to interface with a wide variety of external systems.

- The price-performance balance is *very* attractive. The next step down would be a board based on a low-speed Pentium, i486 or even i386 CPU; these boards are just a few dollars cheaper than the PCM-5820, and much less capable. In particular, the integrated USB is a real boon; it allows you to hook in cheap consumer peripherals rather than fiercely expensive PC/104 expansion cards.

Let's take a moment to examine the PCM-5820 hardware in detail. Figures 2-1 and 2-3 detail photographs of the top and bottom of the board, showing the high level of integration. Note the low-profile heatsink and absence of active cooling. I didn't remove anything from the board to take these photos; this is how the board ships, and it doesn't need any further cooling in most situations.

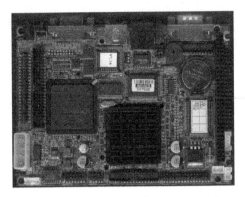

Figure 2-1: Top of PCM-5820   **Figure 2-2: Bottom of PCM-5820**

The major hardware features are as follows:

■ **Microprocessor** – National Semiconductor Geode[6]. The fastest flavor of this processor available on the 5820 is 300 MHz; some other vendors offer 333 MHz products. Geode is a "Pentium-ish" CPU; it is hard to establish an exact equivalent with an Intel CPU, but the performance is something like an accelerated Pentium 1. It supports the MMX-1 instruction set extensions, but it lacks some Pentium core components such as MTRRs (Memory Type Range Registers). Geode has an architectural equivalent, ARRs (Address Range Registers). It also has an extensive system of software traps that allow it to emulate many standard PC hardware features in firmware; more on this topic later. Very roughly speaking, a 300 MHz Geode is comparable in performance to a 200 MHz Pentium with MMX. Archaeologically, Geode is descended directly from the Cyrix MediaGX processor. It appears to share some history with the early IBM/Cyrix 486SLC (clock-multiplied 486-compatible in an i386SX pinout) and "Blue Lightning" (clock-multiplied 486-compatible in an i386DX pinout) processors. Because of the slightly unusual architecture, there are some behavioral oddities in the Geode platform; we'll discuss most of these in the text to follow.

---

[6] The Geode range of x86-compatible Internet appliance processors was sold by National Semiconductor to AMD, in a deal announced in mid-2003. However, as at the time of writing, I have yet to see an AMD-branded Geode chip.

- **RAM** – There is a single standard SODIMM slot supporting memory sizes up to 256 MB. The board uses 3.3V unbuffered PC100 SDRAM. In our examples, we will be assuming a system with 64 MB RAM.

- **Ethernet** – The board has a Realtek RTL8139 10/100 Ethernet MAC; well-supported and relatively trouble-free. There is a network boot extension available in the system BIOS, should you care to use it.

- **USB** – The system has two USB 1.0 OHCI-compatible ports provided by the CS5530 companion IC. I have read scattered reports of problems (lockups and incompatibilities) with the USB implementation in this chip, mostly with high-bandwidth devices (video capture pods, storage devices and LAN adapters). To date, I have not encountered any problems of this nature, and it may be that these issues only affect older operating system kernels.

- **Serial** – There are two serial ports, one of which is RS-232-only, and the other of which can be configured as either RS-232, RS-422 or RS-485.

- **Parallel** – The board features a standard parallel port, configurable for SPP, EPP or ECP modes. This port is very useful as general-purpose I/O.

- **Audio** – The CS5530 companion IC on-board has an AC97 codec interface. At the time of writing, current production of the PCM-5820 is shipping with a Realtek ALC201 codec. Older production used an Analog Devices codec. By and large, this hardware difference should not require any software modifications. The board has line-level and microphone-level inputs, line-level output, and individual speaker drive outputs. It is not capable of delivering much power to the speaker outputs, so for anything other than headphone connections you will probably want an external audio power amplifier.

- **Video** – The 5820 has a standard analog VGA output, as well as a header for connecting to parallel TFT LCDs. An LVDS transmitter IC (and associated LVDS output connector) is optionally available on some board variants. Supported resolutions range up to 1280 × 1024 (at 8 bpp, on CRT only) or 1024 × 768 (at 16 bpp, on CRT or LCD). Passive panels are not supported; the CS5530 requires additional external DRAM to support passive displays, and Advantech has not allocated space on the board for this additional RAM.

- **Mass-storage** – There is a standard floppy drive header supporting two drives. More usefully, there is a single standard IDE bus (with a 44-pin 2 mm pitch "laptop" type connector) and a bootable CompactFlash slot on the secondary IDE port. Note that the CompactFlash slot is wired in True-IDE mode, and it is therefore not possible to use nonstorage devices or to "hot swap" Compact-Flash cards. (The CompactFlash specification requires a power-cycle in order to swap media if the socket is run in True-IDE mode. This requirement has to do with the length of the pins in the socket, which control power sequencing; hot-swap will *sometimes* work on a True-IDE slot, especially if you push the card in swiftly and firmly, but it can't be guaranteed, and you should avoid trying it because there is a risk of damaging the card).

- **Expansion bus** – Although the Geode system uses a PCI architecture, the 5820 does not offer a means to connect PCI peripherals. The board has a standard PC/104 header, essentially an ISA interface.

- **Miscellaneous** – A single PS/2 port allows connection of a keyboard and mouse by way of a Y-cable, supplied with the board. There is also a port to connect an IrDA transceiver or CIR receiver module; the inbuilt IR UART can be configured for various infra-red decoding modes including ASK, FSK and IrDA. (Enabling infra-red functionality usually disables normal use of the second serial port).

If, for whatever reason, you need to seek an alternative supplier of boards, and you're trying to find something similar to the hardware described in this book, there are many options for second-sourcing. (This is yet another advantage of choosing a PC-based architecture). Here is a short list of compatible, or at least broadly similar products from different vendors, with comments on their differences from the PCM-5820. You should be able to run the example code in this book on any of these boards with few or no modifications:

Vendor	Model	Notes
Acrosser *www.acrosser.com.tw*	AR-B1551	Practically identical to the PCM-5820, except for a different mechanical layout and a DiskOnChip socket as well as CompactFlash. The LVDS LCD interface is included as standard on this product. Note that there are a couple of other variants in this family.
BCM[7] *www.bcmcom.com*	EBC-3410	Twin Ethernet ports (based on Realtek RTL8139), otherwise functionally identical to the PCM-5820.
BCM	EBC-5410	This is a 5.25″ form-factor board with four serial ports, a single PCI slot, 64 MB of on-board SDRAM, and a standard DIMM socket for additional SDRAM.
ICP America[8] *www.icpamerica.com*	WAFER-5820	This board has a DiskonChip socket instead of a CompactFlash slot. Otherwise, the product is almost 100% mechanically and electrically identical to the Advantech board, except that the board is not capable of driving loudspeakers directly; it requires an external power amplifier. Note that this same board is sold as a "Gorilla Systems GORWAFER-5820" in some markets.
Netcom IPC *www.netcomipc.com.tw*	NC-529	Very similar to the PCM-5280 except that it has a DiskonChip socket instead of a CompactFlash slot. This board is the "odd man out" of all the other Geode boards I've inspected, in that it uses the National Semiconductor PC97317 Super I/O chip rather than the Winbond W83977AF favored by other vendors. This difference is unlikely to affect you in any significant way, however; the main difference is that the National chip doesn't have quite the same range of infra-red decoding support as the Winbond part.

---

[7] BCM is also known by the brand name e-valuetech.

[8] ICP distributes products from IEI, a Taiwanese OEM. The same products are available from other vendors under different names.

All of the code and other materials in this book have been tested with the PCM-5820, EBC-3410, EBC-5410 and WAFER-5820[9], so if you acquire any one of these boards you can be assured that the examples will run for you "out of the box."

By the way, you should note that although the board outline and screw holes are standardized for the 3.5″ biscuit form factor, the overall mechanical layout is definitely *not* standardized. One example you'll observe in particular is that on the Advantech PCM-5820, the CompactFlash slot is mounted on the solder side of the board, underneath the PC-104 connector. On the BCM EBC-3410 (by way of comparison), the CompactFlash slot is on the solder side of the board, along the same edge as the connector panel. Other important mechanical differences are the layout of connectors on the I/O edge of the board and also the overall airspace requirements of the board, including heatsinks. For instance, the ICP WAFER-5820 has a large custom-made aluminum heat spreader covering both the Geode and CS5530 ICs, and a small, standard-size heatsink is glued on top of that.

The upshot of all this is that you should be aware that it is very difficult to design a completely generic casing that can guaranteeably accommodate all third-party variations on a particular board configuration, unless you're willing to waste a lot of internal space. This is especially true if you need to make connectors on the board directly accessible outside the housing. You should keep this in mind when organizing a product that will have a significant enough production lifespan to require a backup SBC supplier, particularly if your end product needs to meet EMI compliance standards (to earn FCC or CE approval, for instance). It is possible to make your housing fairly generic by cutting a large hole to expose the entire connector edge of the board, but this will increase overall system emissions.

2.6 Selecting an Inter-Module Communications Protocol

When you're building your real-time data acquisition and control systems, you will need to select some kind of interface to connect these peripheral devices with the PC or other "master" system you're using to record and/or analyze the data. Issues you will need to consider when choosing interfaces include:

---

[9] As the WAFER-5820 lacks a CompactFlash slot, obviously I have not tested use of CompactFlash with this board.

- Noise immunity of the selected protocol vs. anticipated noise in your system's environment.

- Data transfer rates and latencies.

- Delivery delays.

- Complexity of any required wiring.

- Maximum permissible cable length (this is usually a function of data transfer rate).

- Cost and difficulty of implementation on the target microcontroller.

- Cost of a matching interface on the PC side, and availability of drivers for the operating system you intend to run on the PC.

- Clock recovery issues, such as maximum allowable system clock drift.

I²C® (Inter-IC Communication), also known as two-wire serial, is a widely-used synchronous serial protocol. It is a half-duplex system implemented on two bidirectional lines, SCL (clock) and SDA (data). Devices on the I²C bus are recognized by means of unique address codes. The issuing authority for these addresses is Philips, which also owns the trademark on the I²C name itself, as well as patents related to its implementation. There are actually three "grades" of I²C: basic (100 kbps, 7-bit addresses), fast mode (400 kbps, 7-bit addresses) and high-speed mode (3.4 Mbps, 10-bit addresses). Faster modes are backwards-compatible with slower modes, and the protocol is designed in such a way that slower peripherals can coexist happily on the same bus with fast devices. Regarding the patent issue, it is not necessary for you to license the interface; ICs that implement I²C include the license cost as part of the chip price. If you study the datasheet carefully, you will see a statement to the effect that the I²C bus implementation is licensed to you with the part, for use with other licensed components. (The reason for the addendum on the end of that statement is to make it clear that using a single licensed component doesn't automatically license everything else on the bus; each individual part needs to have a license).

SPI (Serial Peripheral Interface), also known as three-wire serial, is a mechanically somewhat simpler synchronous serial protocol, the trademark for which is owned by Motorola. Three-wire is a bit of a misnomer, as SPI actually requires four signals per device (plus a ground reference); data in, data out, clock and select. The exact names given to these signals vary among different implementations, but the official names are MOSI (Master Out Slave In), MISO (Master In Slave Out), SCLK and SS (Slave Select), respectively. Note that the data direction in these names (Serial In/Serial Out) is described with reference to the slave device; i.e., MOSI is an output on the master and an input on the slave(s).

SPI works best for single-master, many-slave applications. Because of the need to provide a separate select line to each device that can act as a slave, trying to engineer a system with multiple possible masters is irksome; it's not really what the protocol was designed to do. The advantage of SPI is that it's very simple to implement, it's full-duplex, and it's inherently more efficient than I²C—transfers are initiated simply by asserting the target device's select line, with no additional setup process or addressing handshake phase required before the actual data transfer. The architecture of the interface (from a slave perspective) is simply an 8-bit shift register with the most significant bit connected to MISO and the least significant bit connected to MOSI. While the device is selected, at each clock pulse (polarity is user-definable in most SPI implementations) the shift register rotates left one bit, samples MOSI into its least-significant bit, and the MISO pin is updated with the most-significant bit[10]. If the SS line goes high (inactive), MISO is tristated to prevent bus collisions.

I²C and SPI are frequently used to carry control information around a single board, or between multiple boards in a subassembly; I²C is also frequently used to communicate between a host system (for example, a laptop computer or cellphone) and a "smart" rechargeable battery or other peripheral. You'll also find I²C used variously in consumer A/V equipment (communicating between a microcontroller and tuner, digital potentiometer, display controller and so on) and miscellaneous other appliances (I²C EEPROMs are frequently used to store configuration data in everything from burglar alarms to digital cameras). Neither of these protocols is intrinsically designed to drive long cable runs and both protocols can be both generators of and victims to noise.

---

[10] Note that incoming data is sampled on one clock edge, and outgoing data is latched onto the output pin on the opposite clock edge.

Note also that the official names I²C and SPI are trademarked, and as a result you'll frequently find chip companies implementing very similar, unlicensed interfaces under different names. Such third-party interfaces are usually intentional clones of one or other of these "big two" synchronous protocols; if you're looking at a microcontroller or peripheral that implements some strangely-named synchronous serial interface, the chances are excellent that it is, or at least tries to be, compatible with either I²C or SPI.

One design advantage of synchronous protocols is that clock recovery is intrinsic to the hardware interface, and as long as you don't exceed the maximum permissible data rates, it isn't necessary to maintain tight clock control. This works well, particularly for cost-sensitive applications that use RC oscillators as their clock source. However, there are various reasons you may want to consider one of the standard asynchronous serial interfaces; among which, they are all more amenable to long cable runs.

RS-232—straight asynchronous serial[11]—is the cheap, simple communications standard used successfully in millions of devices for many years. However, complete and correct RS-232 implementations are rarely encountered in consumer-grade electronics such as personal computers, and they are even more rare in embedded devices. Most embedded devices implement one of the following four schemes:

- Simple TTL drivers with a 5 V swing, occasionally biased in some way so that the swing is centered around 0 V. These interfaces are almost always three-wire, that is, they only connect RxD (receive data), TxD (transmit data) and ground. Interfaces of this type are totally out of spec and therefore horribly unreliable. The vagaries of the PC industry are such that some PCs will receive these signals properly (which is why people can get away with designs like this) but many PCs won't work at all. In general, it's a very bad idea to play fast and loose with the standard like this. You'll find this poor man's RS-232 interface used most commonly in hobbyist grade microcontroller programmers (several older PICmicro programmers worked this way, for instance, though mercifully the habit seems to be dying out).

---

[11] It's rarely mentioned, but the RS-232 specification also includes synchronous operation. In practice, virtually no terminal equipment (including PCs) that you'll encounter actually supports synchronous communications, so for all real-world purposes, RS-232 is a purely asynchronous interface. Pedants who assert otherwise are likely to email their complaints in EBCDIC.

- Solutions that use old driver/receiver level shifting chips like the Motorola MC1488 and MC1489, in conjunction with +/–9 V rails (often supplied by back-to-back 9 V batteries, and occasionally supplied by tapping signals on the RS-232 interface itself; the host is relied on to drive those signals to appropriate levels before the peripheral is called upon to function). This kind of interface is dying out, but we still see it from time to time.

- Three-wire charge-pump type driver/receiver implementations using plug-n-play transceiver chips like Maxim's MAX232A. These interfaces usually have a voltage swing between –10 V to +10 V (at least) and are compatible with a wide range of PC hardware.

- Devices which use the previously mentioned charge pump interface chips, and implement at least some of the flow control lines, but fall short of a complete implementation. Probably the most common example of this is to implement RxD, TxD, RTS and CTS. The additional flow control lines are generally not used for their textbook function in embedded devices; they are often used to signal some proprietary status information.

In a few rare cases, peripherals use odd, very proprietary methods to drive the serial lines; one example is hobbyist data slicer circuits for (radio) scanners, which often drive the serial lines directly from the output of an op-amp, the positive and negative rails of which are supplied by two flow control signals from the host. These sorts of systems are mercifully rare. If you're going to use RS-232, I heartily recommend one of the latter two options from the preceding list; if you intend to do high-speed transfers, then flow control is also strongly recommended.

RS-232 is a bidirectional one-to-one communications interface; the standard permits one transmitter and one receiver on each line, and no more. RS-423 is electrically similar (in that it is single-ended; with reference to ground, –4 V to –6 V is defined as mark, and +4 V to +6 V is defined as space) but it is designed for unidirectional, one-to-many communications. RS-423 is rather a rare interface, and I mention it only for completeness. The problems that RS-423 was designed to solve are generally solved even more effectively by RS-485.

RS-422 and RS-485 are differential serial interfaces. These interfaces are capable of driving much longer cable runs (up to 4,000 feet), or higher baud rates (up to 10 Mbps), than RS-232. RS-422 is a multi-drop interface specified to drive up to 10 receivers from a single transmitter; RS-485 is a true multipoint network allowing bidirectional communications amongst up to 32 drivers and 32 receivers on a single two-wire bus. RS-485 is commonly used in applications such as burglar or fire alarm systems, and in industrial control applications.

Note that RS-232, RS-422 and RS-485 driver modes are commonly provided as jumper-selectable options on industrial and commercial single-board computers, so you often get them "free" as part of your system. RS-423 is quite rare and if you want to support it, you will probably have to buy a special converter for your PC.

The great thing about RS-232, RS-422, RS-423 and RS-485 is that it's very easy to test them (all you need is a terminal program), the signals can easily be captured and analyzed on a low-end digital storage oscilloscope (or even, at a pinch, with a piece of software running on your PC), and any operating system will have all the drivers required to talk over these links.

Moving towards the high end of serial protocols, even USB is slowly (and reluctantly) becoming more acceptable as an interface method for embedded systems. Of course, it is already extremely popular in high-volume consumer and commercial applications, but it's much harder to justify selecting it for low-volume or unique systems, simply because there's generally a very large amount of software work (on both the PC and device side) required to get it functional. This ancillary work wastes engineering resources that would be much better spent developing the application of interest. USB is also severely limited as to cable run length, which precludes its use in any application that is not physically adjacent to a PC. Its principal advantage, from the embedded engineer's perspective, is faster transfer speeds than the simplest asynchronous protocols, coupled with reliable hot-pluggability[12] and considerably better noise immunity than the intra-board synchronous protocols described above. In isochronous mode (typically used by USB audio devices) it even has good real-time characteristics. Plus, a welcome side-effect of USB is that it delivers a regulated power supply to your device, although it is quite drastically current-limited (500 mA).

---

[12] Technically, serial and parallel interfaces on PCs are not hot-pluggable. You are *supposed* to power down both the PC and the peripheral to be connected, connect the cable between them, then power up first the PC, then the peripheral.

Most low-volume embedded applications that communicate over USB do so by cheating; they use an off-the-shelf USB interface chip that emulates a standard interface (for example, RS-232) on the device side, and has ready-to-run PC drivers on the PC side. The best-known manufacturer of such chips is Future Technology Devices International Ltd, *http://www.ftdichip.com/*. Although their solutions are about as seamless and plug-n-play as USB development gets, there can still be annoying analog issues to contend with when laying out a PCB using these devices. If you're a true masochist and want to do the device-side USB code as well as write your own driver for the host operating system of interest, probably the most popular parts are Philips PDIUSB011 (serial interface on the microcontroller side) and PDIUSB012 (parallel interface). These chips are readily available from distributors such as Digi-Key.

If you want to go one step further than this, and build your entire embedded app into the USB chip, there are plenty of devices that implement a USB interface. One of the most interesting is Cypress Semiconductor's EZ-USB AN2131QC. This consists of a ROMless 8051 microcontroller with some SRAM and an on-chip USB interface, with an interesting way of getting code into the chip: the driver on the host side downloads the firmware from the PC to the micro, then simulates a detach and reattach event; the micro then attaches itself with its new "personality" determined by the code that was sent to it in the first phase. A very low-cost evaluation board for this chip can be obtained from DeVaSys, *http://www.devasys.com/*. It offers 20 I/O pins and an I²C interface, plus a 16KB EEPROM. (If desired, the micro can be configured to grab its code from the EEPROM instead of relying on the host PC).

For some applications, it may even be useful to employ Ethernet as the communications interface back to the host PC. Although there are numerous protocols that can run over the Ethernet physical layer, for the vast majority of applications, "Ethernet capable" is really a way of saying "runs TCP/IP over Ethernet." The great thing about TCP/IP over Ethernet is that there is a vast selection of ready-made cabling options and traffic forwarding/filtering hardware and software available off the shelf. Provided you implement standard protocols (HTTP, FTP, SNMP and so forth) on the microcontroller end, you also get a free user interface on the PC end in the form of web browsers, SNMP agents, and so on. There are also reference TCP/IP stacks for many microcontrollers. Ethernet is robust, well-understood and reasonably noise-immune, and can (with careful planning) be strung over large distances.

The principal downsides to Ethernet are latency and cost. The really cheap Ethernet parts are of course the high-volume parts used in PC applications; these chips have PCI interfaces and are therefore virtually impossible to interface to small microcontrollers. The market space for embedded Ethernet parts is much smaller. The "gold standard" embedded 10 Mbps Ethernet part is the Crystal Semiconductor CS8900A; another popular choice is the Realtek RTL8019. Both of these parts are in fact standard ISA-bus chips of yesteryear that have been given a reprieve from discontinuation purely because of their popularity in non-PC projects.

Besides the actual cost of the Ethernet MAC chip itself (and a PHY, if applicable), you should also consider the RAM requirements of the TCP/IP stack, the effort required to port a MAC driver and the stack itself to your target architecture, and the difficulty of perfecting all the analog engineering around the Ethernet port. In between the RJ45 jack on your board and the chip that talks to it is room enough for a lot of tedious debugging! For the purposes of this book, I will treat Ethernet capability, if desired, as being one of those functions that's best handled in the soft-task controller.

In addition to these standard interfaces, there are of course numerous proprietary options. For the projects in this text, however I have selected a flavor of SPI. I chose this protocol because it is very simple to implement in firmware, and it is easy to interface directly with a PC. I²C requires bidirectional I/Os on both master and slave device; this adds an extra dimension of complexity when working with PCs, because not all parallel port modes properly support bidirectional I/O. The baseline parallel port specification stipulates that certain signals are inputs and certain signals are outputs; neither reading outputs nor writing inputs is guaranteed to work.

Some Example Sensor, Actuator and Control Applications and Circuits (Hard Tasks)

3.1 Introduction

In this chapter, I will present a few useful "cookbook" applications for real-time control circuits designed to perform some specific low-level task and interface with a master controller for instructions and overmonitoring. For the moment, we will deal principally with the design and firmware of the peripherals themselves. In the next chapter, you will find more detailed detailed explanations of how to develop Linux code to access these peripherals from an embedded PC-compatible SBC or desktop PC. The purpose of this chapter is to provide introductory-level information on how to interface with some common robotics-type sensors and actuators, and in particular to show how these can be tied into the type of system we have been discussing. Although the projects are standalone and don't directly develop on each other, you should read at least the description of the stepper motor controller in full, because that section describes how the SPI slave interface is implemented. This information isn't repeated in the descriptions of the other projects.

Note that in this book, we will discuss an overall system configuration where all devices are connected directly to the Linux SBC, as illustrated in Figure 3-1.

This configuration is easy to develop and test, and is an excellent basis for many types of projects; in fact, this is how I prototyped all the E-2 hardware. For the sake of completeness, however, I should point out that in the actual E-2 system, all of the peripherals are connected to a single master controller (an Atmel ATmega128, in fact). This controller is connected to the SBC over an RS-232 link as illustrated in Figure 3-2.

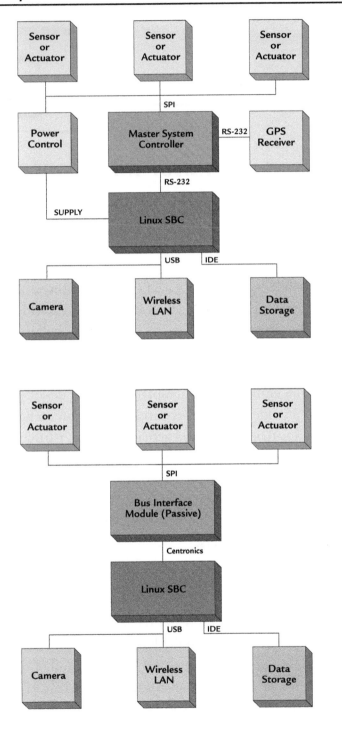

Figure 3-1:
Simplified system layout

Figure 3-2:
Actual E-2 system layout

The master controller is the real brains of the vehicle. In fact, in E-2 the Linux system can be considered just another peripheral of the master controller. The Linux board performs strictly high-level functions; it interfaces to two USB cameras and an 802.11b WLAN adapter, besides writing the vehicle log on a high-capacity storage medium and performing some computationally intensive tasks such as image analysis and digital spectrum analysis of audio coming in from the exterior microphones. This design is basically an engineering refinement of the system we'll be talking about in this book; discussing it in detail really wouldn't add much to the material you already have here. Pay no attention to that man behind the curtain!

For your convenience (and mine, too!), I have developed rough-and-ready PCB artwork for all the example circuits in this book. The PCB artwork is subject to revision, and as a result is not provided on the CD-ROM; you can download it freely from *http://www.zws.com/*. The schematics are, however, provided on the disk. In order to edit the PCB layouts or view the schematics from which they are generated, you will need to install the evaluation version of the Cadsoft Eagle PCB CAD package, which is included on the accompanying CD-ROM. Versions for both Windows and Linux are provided. Please note that these layouts are designed with largely surface-mounted components. This reduces the manufacturing and assembly costs of the PCB (and it also makes routing easier in some circumstances). However, it does make hand-assembly slightly more challenging. The parts I have used can easily be hand-soldered with a little practice, but if you aren't sure of yourself, every part I've used is available in a through-hole version, with the exception of the Analog Devices accelerometer chips.

Ergonomics Tip: A scroll-wheel mouse is highly recommended if you're using Eagle. The wheel controls zoom level. Since the zoom in/out functions are centered on the current position of the mouse cursor, you can navigate all around a large schematic or PCB layout using only the scroll wheel and minimal mouse movements. It's rarely necessary to touch the scroll bars in the Eagle window; it's easier and much faster to zoom out, then zoom back in on the area of interest.

3.2 E2BUS PC-Host Interface

Internal control signals in E-2 are carried on a simple SPI-style ("three-wire") interface[13] using a 10-conductor connector referred to as the "E2BUS" connector. The PCB layouts I have provided with this book use JST's PH series 2mm-pitch disconnectable crimp type connectors. These are commonly used for inter-board connections in applications such as VCRs, printers and CD-ROM drives; they provide fairly good vibration resistance and they hit an excellent price-performance point, as long as you don't mind investing in the appropriate crimp tool. If, however, you are building these circuits on breadboards, you will probably prefer to use standard 5.08 mm (100 mil) headers.

The E2BUS pinout used by the circuits in this book is:

Pin	Name	Function
1	+12 V	+12 VDC regulated supply
2	GND	Ground
3	+5 V	+5 VDC regulated supply
4	GND	Ground
5	MOSI	SPI data input (to peripheral)
6	MISO	SPI data output (from peripheral)
7	SCK	SPI clock
8	_SSEL	Active low slave device select line
9	_RESET	Active low reset input
10	GND	Ground

E2BUS is specified to carry up to 500 mA on each of the 12 V and 5 V lines. Peripherals that expect to draw more than 500 mA on either rail should have separate power input connectors (the main drive motor controller is one example that falls into this category).

---

[13] Note that 3-wire SPI is in no way related to "three-wire serial" RS-232 interfaces, which are simply a normal serial connection with only RxD, TxD and ground connected. SPI is a synchronous protocol.

There are two useful things to note about the E2BUS connector:

1. It's possible to assemble a cable that will let you connect a PC's parallel port directly to an E2BUS peripheral (at a pinch, you can dispense with buffering and simply run wires direct from the parallel port signals to the E2BUS device). A fairly simple bit-banging piece of software on the PC will allow you to communicate with the peripheral.

2. The E2BUS interface brings out all the signals necessary to perform in-system reprogramming of the flash and EEPROM memory of the AVR microcontrollers we are using, so in theory this port could be used to update the code and nonvolatile parameter data, if any, in an E2BUS module without needing to remove the microcontroller. For various reasons, however, it isn't always possible to achieve this with an AVR-based circuit; either because the ISP pins are being used for other functions by the circuit, or because the microcontroller lacks an external clock source (which may be required for in-system programming). However, the connector design is, at least, flexible enough to allow the possibility if you want to take advantage of it.

At this point, you might be wondering why I chose to use SPI rather than, say, I²C (which requires fewer I/O lines and would allow a true "bus" configuration with a single set of signals routed to all peripherals) or CAN, which is better suited for unfriendly environments such as automotive applications. The first reason is code simplicity. CAN and I²C are both, by comparison with SPI, relatively complex protocols. For example, I²C uses bidirectional I/O lines and it's a little complicated to isolate an I²C device from the rest of the bus, because your isolation component needs to understand the state of the bus. I²C is also best suited for applications where a master device is programming registers or memory locations in a slave device. SPI is a slightly better protocol—with virtually no overhead—for peripherals that deliver a constant stream of data.

For the purposes of this book, we'll primarily be talking about controlling E2BUS peripherals directly from the parallel (Centronics) printer port of a PC-compatible running Linux. This is the easiest scenario to describe, and it illustrates all of the required techniques nicely. Following is a schematic for a fairly simple parallel port interface that allows you to connect up to eight SPI-style peripherals to a PC. The schematic for this project is available in the projects/parbus directory on the CD-ROM. By means of LEDs, the interface shows you which device is currently selected, and activity on the data input and output pins.

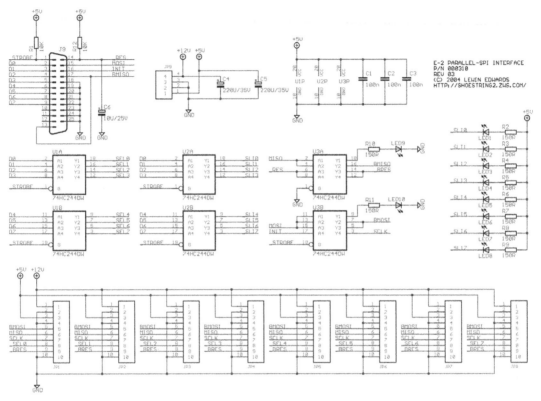

Figure 3-3: Parallel port E2BUS interface

This circuit might appear unnecessarily complicated, but it's really quite simple. The eight data lines from the parallel port are used as select lines for the eight peripherals. These signals are buffered through 74HC244s, the outputs of which are tristated by the parallel port's _STROBE signal. The reason for the tristate control is to reduce the chance of spurious bus transactions while the SBC is performing power-on initialization. NOTE that this system assumes that the device(s) in use in your peripherals have their own pullup resistors on the select lines. An additional HC244 buffers the same signals to a row of indicator LEDs that show you which device is currently selected. A third HC244 buffers the control signals used for MISO, MOSI and SCK, and additionally drives the _RESET line.

A side benefit of this circuit: If you use 5 V-tolerant, 3.3 V-capable devices where I have specified 74HC244s, you can use the design in Figure 3-3, virtually unmodified, to communicate between a standard 5V-level PC parallel port and external devices that use 3.3 V I/Os.

If you're looking at the schematic I provided on the CD-ROM, you'll observe that my accompanying PCB layout includes a standard right-angle DB25M connector to mate directly with the parallel port on a PC. If you are planning to build some kind of enclosure containing an SBC and connected E2BUS-style peripherals, you might instead consider using a 26-pin, 2 mm or 0.1"—spaced header. Most SBCs use one or other of these connectors for their parallel port.

In fact, you don't need to build this entire circuit to communicate with the projects in this book. If you *only* want to talk to one peripheral at a time, if you're exceedingly lazy, and if you're willing to take a bit of a risk on port compatibility, you can experiment with a quick-n-dirty cable wired as follows. The left-hand column indicates the E2BUS pin number, and the right-hand number indicates which corresponding signal should be wired on a DB25M connector.

Pin	Name	Connect to
1	+12 V	External +12 VDC regulated supply
2	GND	+12 VDC ground return
3	+5 V	External +5 VDC regulated supply
4	GND	+5 VDC ground return
5	MOSI	Pin 15 of DB25M.
6	MISO	Pins 17 and 13 of DB25M.
7	SCK	Pin 16 of DB25M.
8	_SSEL	Pin 2 of DB25M.
9	_RESET	Pin 14 of DB25M.
10	GND	Ground, pins 18–25 of DB25M.

Be warned—there is **absolutely no protection** for your computer's parallel port if you use this circuit. If you accidentally short, say, a 24 V motor supply onto one of the parallel lines, you will need a new motherboard. I strongly warn you not to use this quick and dirty hack with a laptop computer, unless it's a disposable $50 laptop you bought off eBay!

Also be warned that the simple cable is substantially less tolerant of variations in the motherboard's parallel port implementation than the full E2BUS interface board. If you find yourself missing transmit or receive bits, or getting garbage data, try adding a rather strong pullup, say 1K, to the SCLK and MOSI lines. If you still have problems, it may be possible to mitigate them by slowing down your data rates, but there will certainly be some trial and error waiting for you.

As I mentioned in the introduction to this chapter, the actual E-2 project isn't structured exactly as I have described in this section, and the principal reason is energy consumption. The PCM-5820 and its dependent peripherals are the greediest power hog in the entire submarine (these modules of the circuit pull considerably more current than both drive motors operating at full speed), and its brains aren't required most of the time on a typical E-2 voyage. For this reason, the master controller on the voyage is another AVR microcontroller— an ATmega128, to be exact. The peripheral select signals are generated by three GPIOs fed to a 74HC138 1-of-8 decoder. However, I originally started the project by connecting the peripherals directly to the SBC in the manner described in Figure 3-1, because it was the easiest way to debug the protocol and the peripherals themselves. For an early prototype, or for any laboratory fixture application that doesn't require battery power, you almost certainly want to do the same thing; it's much less challenging to debug the protocol and front-end interface issues in this configuration.

In the interest of completeness, I should point out one major weakness of the simplified E2BUS design in this book: It relies on the peripherals to perform bus arbitration. The ATtiny26L doesn't implement a full SPI interface in hardware, so the firmware in each peripheral needs to track the state of the select line and manually tristate its serial data output line when deselected. If any module happens to crash in an on-bus state, the entire bus could potentially be brought down. This design flaw could be mitigated to some degree by adding tristate buffers gated by the select line, or by migrating the peripherals to a different microcontroller that implements the full SPI interface in hardware. Also observe carefully that there is no reset generation circuitry on the individual peripheral modules; they rely on receiving an explicit software-generated reset from the attached SBC. A real-world design should implement an external reset generator with brownout detection, to ensure that all modules are reliably reset after a brownout or power-up event.

3.3 Host-to-Module Communications Protocol

The SPI specification only defines the bare outline of the communications proto-col, including little more than the physical interface. This is a good thing and a bad thing. It's good, because you can make your protocols as simple as you like— and bad, because it means you have to specify and develop your own high-level protocols! The basic rules are as follows: Each slave device has an active-low slave select line (SS), a clock input (SCK), a data input (MOSI) and a data output (MISO). Note that the words "input" and "output" here are with reference to the *slave* device. It is fairly normal practice in schematics of SPI equipment to label the entire "output to slave(s)" net as MOSI and the "input from slave(s)" net as MISO, which can be slightly, and pulse SCK high. At this point we can sample the data stream out of the micro at MISO. Here's a sample waveform where the host is sending the code 0xFE to a peripheral. The top trace is MOSI and the bottom trace is SCK. Note how the pulses have rounded leading edges ("shark fins"). This trace was captured on a system connected using the quick and dirty cable as described previously.

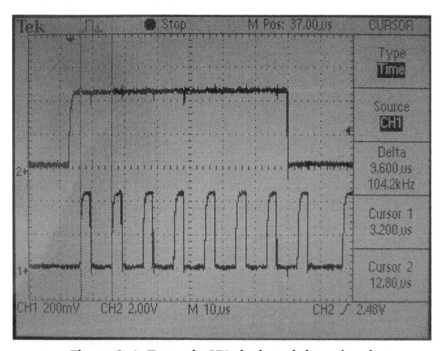

Figure 3-4: Example SPI clock and data signals

The bit cell is approximately 9.6 µs, corresponding to a serial clock rate of 104.2 kHz. This is the fastest speed we can get out of the PCM-5820 using the code in e2bus.c with all timing delays commented out. Note that we're only using half the available bandwidth; it's entirely possible to implement a full-duplex protocol over the interface described in this section.

From a design perspective, you should observe also that for the projects described here, the Linux machine is always the bus master. This is a significant weak point in system reliability, because a crashed Linux box could potentially leave one or more peripheral modules in the "selected" state, listening to random noise coming down the bus. If you plan to implement a real system with this architecture, you should implement hardware and/or firmware interlocks to prevent such occurrences. For example, you could implement a timeout in the routine that monitors the SS line; if there is no SCK within a specified time period from SS going active, the peripheral should assume a crashed master, and go off-bus. Of course, this doesn't help you if the Linux box has pulled the master reset line low. You shouldn't use a configuration like this to control hardware that may need to be "safed" in event of a loss of control, unless you have some other external hardware that can overmonitor the control system and shut things down gracefully if the controller fails.

I have developed a simple piece of Linux code to do all the synchronous serial I/O you will need to talk to these projects. This code is provided in the projects/e2bus directory on the CD-ROM. The meat of this code resides in five simple C functions. Note that these functions assume that your E2BUS interface is connected on the first parallel port. Also note that the timing they exhibit is quite sloppy, since we're not attempting to make Linux appear real time. You should not run this code inside a low-priority thread, because other things will preempt it and may cause spurious timeout problems.

Following are the basic function prototypes:

Prototype	Description
int E2B_Acquire(void)	You must call this function before calling any other E2BUS functions. It attempts to get exclusive access to the first parallel port. It returns 0 for success or −1 for any error.
void E2B_Release(void)	You can call this function as part of your at-exit cleanup routines. It ensures that all devices are deselected, and releases the parallel port. If you exit without calling this function, the port will still be released implicitly as your task ceases to exist, but devices may still be selected.
void E2B_Reset(void)	Deselects all devices, asserts the reset line on the SPI bus for 250 ms, then pauses for an additional 250 ms before returning.
void E2B_Tx_Bytes(unsigned char *bytes*, int *count*, int *device*, int *deselect-after*)	Asserts the select line for the specified *device* (valid device numbers are 0–7), then clocks out the specified number of bytes one bit at a time. If *deselect-after* is nonzero, the device is deselected after the transmit operation is complete. Setting this argument to 0 allows you to read back a command response without having to set up a new SPI transaction.
void E2B_Rx_Bytes(unsigned char *bytes*, int *count*, int *device*, int *deselect-after*)	Works exactly the same as E2B_Tx_Bytes(), but receives data instead of transmitting it.

These functions, particularly E2B_Rx_Bytes and E2B_Tx_Bytes, are the low-level underpinnings of the E2BUS protocol. The workings of these functions are described in more detail, along with the complete sourcecode, in Section 4-6.

On the device end, all the example circuits here share pretty much exactly the same code for serial transfer operations, though command processing details are naturally specific to each project. Incoming SPI data is received by the ATtiny26L's USART and processed by a very simple and hence robust state machine. You'll find the states defined at the start of the sourcecode for each project, with constants named FSM_xxxx. When a device's SEL line is inactive, the state machine is in a quiescent mode (FSM_SLEEP); the MISO pin is set to input mode (to prevent it from driving the bus); clock and data from the USI are ignored, and USI interrupts are disabled. Asserting SEL pushes the state machine into a "listen for command byte" mode, resets the USI, and enables data receive interrupts. The first complete byte received generates an interrupt which causes a state transition. The destination state is determined by the value of the command byte received. The machine may transit through further states depending on whether the command requires additional data bytes or not. If the received command requires additional data, the system proceeds through intermediate states to receive these additional byte(s), and then executes the command before returning to quiescent mode.

If the destination state involves transmitting data back to the host, the data required for transmission is assembled for return to the host, and subsequent USART overflow (or rather, underflow) interrupts clock the data out a byte at a time. After the last reply byte is clocked out, the final underflow interrupt causes a transition back to the quiescent state.

Deasserting SEL at any time immediately disables the USART and tristates MISO. This completely aborts any data transfer or command in progress; any partially received command will be discarded, and partially-transmitted data blocks will be forgotten.

3.4 Stepper Motor Controller

Stepper motors are useful for relatively low-speed, intermediate-torque drive and positioning applications, particularly where accurate sub-revolution rotor position control is necessary. Motors of this type are commonly used to drive the reels on electromechanical slot machines (one-armed bandits), to position floppy disk drive heads, operate trainable camera platforms, and to power the drive wheels of small mobile robots. In times of yore, they were also used to position hard disk heads, though such applications have long ago been taken over by voice-coil type mechanisms. Stepper motors are simple and cheap to use, and you don't need to have a fully closed-loop controller to use them accurately. Servomotors are much faster, but for guaranteeable positioning accuracy, you need to have a position encoder on the shaft to provide feedback on the actuator's position. By contrast, as long as you don't stray outside your system's nominal acceleration profile (see the following), a stepper-based system can reliably maintain its position indefinitely without recalibration.

There are several types of stepper motor, with varying electrical drive requirements. However, by far the most common type of motor to be found on the surplus market (or scavenged from unwanted computer equipment) is the four-pole unipolar type[14], so this is the type our circuit is designed to use. Without further ado, here's the schematic[15]:

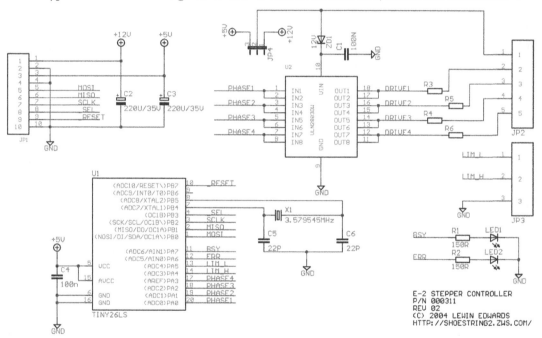

Figure 3-5: Stepper motor control circuit

---

[14] When faced with an unknown stepper motor of small to intermediate size, a *very* reliable gamble to play is as follows: if it has more than four wires, it's probably a four-pole unipolar motor, 0.9 degrees per step, and likely rated for either 5 V or 12 V operation. There are vast numbers of motors constructed with these characteristics.

[15] The alternate function for pin 1 is misprinted here as NOSI—it should be MOSI. This is an unimportant typographical error in the atmel.lbr library supplied by Cadsoft as part of the EAGLE package.

This project uses the ULN2803 octal high-voltage, high-current Darlington array to switch the stepper coils. This chip is readily available for around $0.75 in small quantities, and it is a handy solution for driving moderate loads. Until recently, one could often find this chip, or its close relatives, in commercial stepper motor applications such as inkjet printers and both sheet-fed and flatbed scanners. At present, however, it appears to be in decline as application-specific microcontrollers with high-current drivers on-chip take over its market space. On the subject of prices, you'll notice that I've specified an NTSC colorburst crystal as the clock source, despite the fact that the tiny26L is rated at up to 8 MHz for a 5 V supply voltage. I chose the 3.579545 MHz value, although it's not a nice integer to work with, because these crystals are available everywhere and are often cheaper than other speeds. Chances are you have several in your junkbox already, in fact. You'll also find that application notes for microcontrollers almost always give precalculated example timing constant values (e.g., for setting the baud rate of a UART) for this base clock speed.

Our example stepper controller module also has two active-low limit switch inputs. These are optionally used to signal end-of-travel in the increment and decrement step directions. Note that JP4, which selects between 5 V or 12 V drive for the stepper coils, is intended to be a wire link for factory configuration, rather than a user-changeable jumper. If you are using the device in 5 V drive mode, you should alter or remove ZD1; you can also omit C2, since it serves no function if you're driving the motor off the +5 V rail.

The controller operates in one of two modes: "drive" or "train." In drive mode, you simply specify a speed and direction, and the motor turns in that direction until commanded to stop. Optionally, you can request that it travel until either of the limit switches is triggered. Train mode is intended for positioning applications. In this mode, you command the stepper controller to seek to a specific offset from the current position, and it will automatically seek to that position while you carry out other tasks. The stepper will automatically cut off if it hits the high limit switch while seeking forwards, or the low limit switch while seeking backwards.

Note that the limit switches are permanently associated with specific seek directions. The "low" limit switch is only enforced for "backwards" seeking, and the "high" limit switch is only enforced for "forwards" seeking. The reasons for this are twofold: First, an external force—say, water rushing past a submarine's rudder—might turn the stepper past the make-point for the limit switch, before it reaches a mechanical stop. Second, switches are practically never perfect—in other words, the displacement required to make a contact isn't necessarily the displacement required to break it. You might need to push the arm of a microswitch two steps in to penetrate the oxide layer on its contacts; the first step in the other direction might leave the cleaned metal contact surfaces still touching. Or you might be using a reed switch—you need to bring the magnet to a certain proximity to close the switch, but a weaker field will suffice to hold the switch closed. In any of these sorts of cases, it could require one or more "extra" reverse steps to clear the limit condition.

The stepper controller accepts 8-bit command bytes, optionally followed by additional data. Essentially the same serial reception code is used in all the projects in this book, so it deserves a little additional study here. To begin with, please note that my choice of I/O pin assignments was by no means arbitrary. The AVR's pin-state-change interrupts are useful, but not very intelligent. On the tiny26L, there are only two such interrupts: PCINT0, which (if enabled) fires on state-changes for pins PB0-PB3, and PCINT1, which fires on state-changes for pins PA3, PA6, PA7, and PB4-PB7. When one of these interrupts fires, there is no direct way of determining which pin caused the interrupt; you have to maintain a shadow copy of the port registers and compare them to determine which pin(s) changed state.

Fortunately, when an alternate function is enabled for a pin, that pin will no longer generate state-change interrupts (note that there are a couple of exceptions to this rule). Even more fortunately, the three USI signals used for SPI-style communications are mapped to pins PB0-PB2. Thus, by configuring the USI in three-wire mode, PCINT0 will fire only if PB3 changes state. Since the USI in the tiny26L doesn't implement slave select logic in hardware, we need to do it in software—and as a result of all the discussion in the previous paragraph, it makes excellent sense to use PB3 as the SPI select line, since it has a state-change interrupt all to itself.

The entire meat of the stepper code is contained in three interrupt handlers: USI overflow, timer 0 overflow, and PCINT0. PCINT0 is probably the single most important function in the firmware—it is responsible for checking the state of PB3 and disabling the output driver on MISO (PB1) when the stepper controller is deselected (so we don't fight with anything else on the bus), or enabling it if _SSEL is asserted.

When the device is deselected, this ISR also disables USI interrupts, because we don't care about other transactions that may be occurring on the bus, and having to service USI interrupts causes timing jitter in any step operation we happen to be running in the background. Here's the code in this handler:

```
;===================================================================
; I/O pin change interrupt
; The only valid source of this interrupt is PB3, which is used as
; the 3-wire slave select line.
entry_iopins:
    push r0
        push r16
        push r17
        in r0, SREG

        ; Check state of select line, which is the only line that should
        ; have generated this interrupt.
        sbic PINB, PORTB_SEL
        rjmp usi_disable

        ; SEL line is LOW. Enable and reset USI and switch PB1 to output
        ldi r24, FSMS_RXCMD
        ldi r16, $00
        out USIDR, r16        ; Empty USI data register
        out USISR, r16        ; Clear USI status (including clock count!)
        sbi DDRB, PORTB_DO    ; set PB1 to output

        sbi USISR, USISIF     ; Clear start condition status
        sbi USISR, USIOIF     ; Clear overflow status
        sbi USICR, USIOIE     ; Enable USI overflow interrupts

        rjmp iopin_exit

        ; SEL line is HIGH. Disable USI and switch PB1 to input to take
        ; us off-bus
usi_disable:

        ; disable USI start and overflow interrupts
        cbi USICR, USISIE
        cbi USICR, USIOIE

        ; Disable output driver on PB1 (DO)
        cbi DDRB, PORTB_DO    ; set PB1 to input
```

```
iopin_exit:

        out SREG, r0
        pop r17
        pop r16
        pop r0
        reti
```

Actual stepping operations are performed in the timer 0 overflow interrupt. Timer 0, which has an 8-bit count register, is clocked through the prescaler at CK/256, which is approximately 14.053 kHz. When the overflow interrupt fires, the first thing the handler does is to reload the timer register with a step speed value. The default speed value is $00. Since timer 0 counts upwards, this means that by default the step speed is roughly 55 Hz, which is the slowest configurable speed. You can configure faster speeds by using the CMD_STEP_SETTICK command, followed by an 8-bit parameter that sets a new (larger) reload value. For instance, if you configure a reload value of $E0, the timer will overflow every 33rd ($21) tick instead of every 256th, thereby yielding a step speed of approximately 425 Hz. Theoretically, you could specify a reload value of $FF, resulting in an overflow on every tick and a 14.053 kHz step speed, but in practice there is an upper boundary on legal values for the timer reload figure. This boundary is set by the number of CPU instruction cycles required to service an incoming interrupt and make ready for the next, and it caps the step speed at about 7.1 kHz (reload value $FE) for the cheap NTSC colorburst clock crystal I specified. This shouldn't be a serious impediment: although many stepper motors are rated for as much as 10,000 steps/sec, real applications rarely exceed 2,000 steps/sec (300rpm) due to the fact that the torque of a stepper motor rapidly decreases as step speed increases. The code for Timer 0 handler, along with the subroutines it calls, is as follows:

```
;=======================================================================
; Timer 0 overflow
entry_timer0:
        push r0
        push r16
        push r17
        in r0, SREG

        ; Reset TMR0 counter to start position for next tick
```

```
        lds r16, tick_speed
        out TCNT0, r16

        ; Update state of limit switch flags in machine
        ; status byte (for the benefit of the main thread only)
        sbr r25, (1 << LIM_H)
        sbic PINA, PORTA_LIM_H
        cbr r25, (1 << LIM_H)

        sbr r25, (1 << LIM_L)
        sbic PINA, PORTA_LIM_L
        cbr r25, (1 << LIM_L)

        ; Load current tick-command and see what we should be doing
        cpi r19, TICK_FWD
        breq tick_seek_fwd
        cpi r19, TICK_REV
        breq tick_seek_rev
        cpi r19, TICK_POWERDOWN
        breq tick_poweroff
        cpi r19, TICK_SEEK
        breq tick_seekto
        cpi r19, TICK_SEEK_FWEND
        breq tick_seek_fwdend
        cpi r19, TICK_SEEK_RVEND
        breq tick_seek_revend

        ; Note - TICK_SLEEP falls through to here
        rjmp tick_done

;======================================================================
; Tick event - Seek forward, ignoring limit switch
tick_seek_fwd:
        rcall seek_fwd
        rjmp tick_done

;======================================================================
; Tick event - Seek backward, ignoring limit switch
tick_seek_rev:
        rcall seek_rev
        rjmp tick_done
```

```
;=====================================================================
; Tick event - Seek forward, honoring limit switch
tick_seek_fwdend:
      sbis PINA, PORTA_LIM_H
      rjmp seekto_finished

      rcall seek_fwd
      rjmp tick_done


;=====================================================================
; Tick event - Seek backward, honoring limit switch
tick_seek_revend:
      sbis PINA, PORTA_LIM_L
      rjmp seekto_finished

      rcall seek_rev
      rjmp tick_done


;=====================================================================
; Tick event - Power down motor
tick_poweroff:
      andi r25, ~(1 << SEEKING)          ; Turn off busy flag
      in r16, PORTA
      andi r16, $F0
      out PORTA, r16
      ldi r19, TICK_SLEEP
      rjmp tick_done


;=====================================================================
; Tick event - Generic seek operation
tick_seekto:

      ; First check if the step count is 0 - if it is, then there's
      ; nothing left to do and we should go back to sleep.
      cpi r23, $00
      brne seekto_nz
      cpi r22, $00
      brne seekto_nz
      cpi r21, $00
      brne seekto_nz
      cpi r20, $00
      brne seekto_nz

      rjmp seekto_finished
```

```
seekto_nz:
      sbrc r25, DIRECTION
      rjmp seekto_fwd

      ; Seekto - REVERSE
      ; Check limit switch. If it's active, we stop.
      sbis PINA, PORTA_LIM_L              ; Check limit switch
      rjmp seekto_finished

      rcall seek_rev
      rjmp seekto_update_count

      ; Seekto - FORWARD

seekto_fwd:
      sbis PINA, PORTA_LIM_H              ; Check limit switch
      rjmp seekto_finished

      rcall seek_fwd
      rjmp seekto_update_count

seekto_update_count:
      dec r23
      cpi r23, $FF
      brne seekto_notz
      dec r22
      cpi r22, $FF
      brne seekto_notz
      dec r21
      cpi r21, $FF
      brne seekto_notz
      dec r20

   ; No terminal conditions have been encountered - continue stepping
seekto_notz:
      rjmp tick_done

seekto_finished:
      ldi r19, TICK_SLEEP
      andi r25, ~(1 << SEEKING)          ; Turn off busy flag
      rjmp tick_done
```

```
tick_done:
        out SREG, r0
        pop r17
        pop r16
        pop r0
        reti


;====================================================================
; SUBROUTINE - Seek forward one step
; Destroys R16, R17, SREG
; Updates X,Y
; Implicitly powers up motor and leaves it in powered state
seek_fwd:
        lds r16, stepper_phase
        inc r16
        andi r16, $03              ; Only the lower two bits interest us
        sts stepper_phase, r16     ; Store new current phase

        ldi r17, $01

sf_lp:
        cpi r16, $00
        breq sf_lp_done
        dec r16
        lsl r17
        rjmp sf_lp

sf_lp_done:
        in r16, PORTA
        andi r16, $F0
        or r16, r17
        out PORTA, r16

sf_update:
        inc r29                    ; increment step position
        brne sf_done
        inc r28
        brne sf_done
        inc r27
        brne sf_done
        inc r26

sf_done:
        ret
```

```
;=======================================================================
; SUBROUTINE - Seek backward one step
; Destroys R16, SREG
; Updates X,Y
; Implicitly powers up motor and leaves it in powered state
seek_rev:
        lds r16, stepper_phase
        dec r16
        andi r16, $03          ; Only the lower two bits interest us
        sts stepper_phase, r16  ; Store new current phase

        ldi r17, $01

sr_lp:
        cpi r16, $00
        breq sr_lp_done
        dec r16
        lsl r17
        rjmp sr_lp

sr_lp_done:
        in r16, PORTA
        andi r16, $F0
        or r16, r17
        out PORTA, r16

sr_update:
        ; Check for zero condition.
        sbic PINA, PORTA_LIM_L
        rjmp sb_notz
        clr r26
        clr r27
        clr r28
        clr r29
        ret

sb_notz:
        dec r29                    ; decrement step position
        cpi r29, $FF
        brne sr_done
        dec r28
        cpi r28, $FF
        brne sr_done
        dec r27
        cpi r27, $FF
```

```
        brne sr_done
        dec r26
        cpi r26, $FF
        brne sr_done

sr_done:
        ret
```

The most complex code segment, at least in terms of code volume, is the USI handler. This handler implements a simple state machine. The first byte received after the select line goes active is a command byte. This byte either causes the USI receive ISR to modify the system state directly, or to transit the ISR's state machine through further intermediate states either to receive more data, or to transmit a multi-byte response back to the host. I won't reproduce the code here, because it's largely a very long switch ... case statement implemented in assembly language.

The theoretical transfer speed between the stepper controller and its master is limited by a couple of factors. First, hardware absolutely limits the USI to transfer clock rates of $f_{CK}/4$ in three-wire slave mode. In our case, this is approximately 895 kHz, and we have to be careful that the SPI master doesn't exceed this speed. In the case of a PC parallel port master, as we are discussing in this book, it is unlikely (however, not entirely impossible) that you will be able to outrun the USI. The reason for this is tied up in the ancient PC architecture of twenty years ago, and the fact that the parallel interface is designed for backwards compatibility with 9-pin dot-matrix printers run by slow 8-bit microcontrollers. Because of these compatibility issues, the parallel port registers are, conceptually or physically, on the other end of an ISA bridge; they are limited to ISA bus clock speeds (nominally 8 MHz) and may even have additional wait states inserted. Furthermore, the layers we traverse when making calls to change the parallel port registers add extra delays, the constant SMM interrupts on the Geode platform are stealing cycles from us regularly—and to cap it all off, the application layer in Linux is inherently so nonreal-time as to make the idea of accurately marking off 1 μS delay steps in userland quite silly indeed. You will observe, therefore, that the SPI code I provide in e2bus.c does not have explicit timing instructions throughout all of the state-changes. This code is tested on the 300 MHz Geode platform with Advantech's BIOS 1.23; if you need to rely on it to work faultlessly, it will need, at minimum, testing and requalification on other systems.

Following is a complete list of commands recognized by the stepper controller's firmware. (The mnemonic names are defined for you in e2bus.h):

Mnemonic	Value	Description
STEP_CMD_STOP	0	Aborts any current step operation. Note that the current step position, and the steps-remaining counter, are not altered—you can resume the step operation later, if necessary. The stepper motor coils remain powered.
STEP_CMD_SLEEP	1	Aborts the current step operation and de-energizes all stepper coils. The motor may move a step or two unpredictably when you re-energize it; a calibration may be necessary.
STEP_CMD_SETTICK	2	Set the step rate (larger values = faster rate). This command should be followed immediately by a one-byte step rate.
STEP_CMD_DRIVE_FWD	3	Starts the motor driving forwards at the configured step speed. The motor will continue to spin until another step command is received, ignoring the limit switches.
STEP_CMD_DRIVE_REV	4	Works the same as STEP_CMD_DRIVE_FWD but steps backwards instead of forwards.
STEP_CMD_STEP_FWD	5	Steps forwards a specified number of steps, or until the "high" limit switch is closed. This command should be followed immediately by a four-byte step count (MSB first).
STEP_CMD_STEP_REV	6	Works the same as STEP_CMD_STEP_FWD but steps backwards instead of forwards, and monitors the "low" limit switch. The current step position is set to 00000000 when this switch is closed.
STEP_CMD_FWD_END	7	Steps forwards continuously until the "high" limit switch is asserted.

Mnemonic	Value	Description
STEP_CMD_REV_END	8	Steps backwards continuously until the "low" limit switch is asserted. The current step position is set to 00000000 when this switch is closed.
STEP_CMD_READ_STATUS	254	Reads back the controller status byte and four-byte step position.
STEP_CMD_RESET	255	Performs a soft reset.

The status byte returned by STEP_CMD_READ_STATUS is formatted as follows:

Bit	Function
7	Unexpected interrupt or internal error detected.
6	Step error, e.g., attempt to seek beyond calibrated range. (LED2 – ERR – tracks the state of this bit).
5	Reserved.
4	Reserved.
3	Reserved.
2	High limit switch asserted.
1	Low limit switch asserted.
0	Currently seeking to requested position (LED1 – BSY – tracks the state of this bit).

This is an *extremely* simple stepper control design; it is intended for low-speed positioning and simple low-torque drive applications only. The E-2 project uses these modules to position its rudder and dive planes, and to swivel a camera platform, neither of which are particularly demanding applications. For a little more understanding of the subtleties of stepper motor control, try running this little code snippet, which assumes that you have a stepper module connected as E2BUS device #0. (You'll find this sourcecode, with a makefile, in the stepper1 directory of the sample sourcecode archive):

```
#include <stdio.h>
#include "e2bus.h"
int main (int _argc, char *_argv[]) {
      unsigned char pkt[2];

      // Open port and start stepper motor
      if (E2B_Acquire()) {
            printf("Error opening E2BUS.\n");
            return -1;
      }

      E2B_Reset();

      pkt[0] = STEP_CMD_DRIVE_FWD;
      E2B_Tx_Bytes(pkt, 1, 0, 1);

      // Speed motor up gradually
      pkt[0] = STEP_CMD_SETTICK;
      for (pkt[1] = 0; pkt[1]<255; pkt[1]++) {
            printf("Setting speed: %d\n", pkt[1]);
            E2B_Tx_Bytes(pkt, 2, 0, 1);
            sleep(1);
      }

      // Stop motor and de-energize it
      pkt[0] = STEP_CMD_SLEEP;
      E2B_Tx_Bytes(pkt, 1, 0, 1);

      return 0;
}
```

This will take a few minutes to complete its run. While it's proceeding, listen to and watch your motor. You will observe two things:

1. Certain step speeds are very noisy, but there will be a range of speeds—typically, the faster speeds—for which the motor is comparatively silent.

2. Depending on your stepper motor, at some point on the speed ramp, the motor will probably stop spinning and will simply begin to hum. Take a note of the approximate speed value when this happens; for the motors I am using (under no mechanical load), this is about 240.

The second point in particular is important, and needs elucidation. Try running this second code snippet (this project is located in the stepper2 directory):

```
#include <stdio.h>
#include "e2bus.h"

int main (int _argc, char *_argv[])
{
        unsigned char pkt[2];

        // Open port and start stepper motor
        if (E2B_Acquire()) {
                printf("Error opening E2BUS.\n");
        }
        E2B_Reset();

        // Set the maximum speed your motor can withstand
        pkt[0] = STEP_CMD_SETTICK;
        pkt[1] = 230;
        printf("Setting speed: %d\n", pkt[1]);
        E2B_Tx_Bytes(pkt, 2, 0, 1);

        printf("Starting motor.\n");
        pkt[0] = STEP_CMD_DRIVE_FWD;
        E2B_Tx_Bytes(pkt, 1, 0, 1);

        sleep(2);

        // Stop motor and de-energize it
        pkt[0] = STEP_CMD_SLEEP;
        E2B_Tx_Bytes(pkt, 1, 0, 1);

        printf("Finished.\n");

        return 0;
}
```

Here we attempt to set a speed just shy of the fastest possible step rate we observed in the first test, while the motor is stopped—and then we try to start the stepper. You will observe, however, that it doesn't rotate at all—it just stalls and whines like an engineer awakened by his wife's alarm clock. (If you don't see this behavior, then increase the speed value a little, recompile, and try again).

To understand what's going on here, you need to think about the mechanics of the situation. Each position of the stepper's rotor is a stable mechanical state for a certain corresponding electrical (magnetic) state of the coils. This state can be visualized as a valley between two hills of a plot of net force vs. angle; the motor tries to seek the low point in the valley, where the clockwise and counterclockwise forces on the rotor are equal. If you put an external mechanical force on the rotor, in (say) a clockwise direction, you are pushing up one side of the hill. If you push all the way to the top, then when you release the rotor, it will fall down into the next valley, i.e., the next stable step position.

When we advance the step phase in software, we alter the electrical state of the coils, which creates a new position of mechanical equilibrium. Effectively, we move the two hills and valley along a quarter-phase, which means the motor is no longer sitting in the middle of the valley. Since the clockwise and counterclockwise forces on the rotor are no longer equal, the rotor turns until it is in an equilibrium position once again; the step operation is then complete. However, this process doesn't happen instantly, because the rotor – and whatever mechanical load it's driving—has inertia. The time required for the rotor to find its new equilibrium point depends, among other things, on the inertia (which is determined by the mechanical load on the motor) and the force exerted by the coils—which is directly proportional to coil current.

The above explains both why we observe unreasonable stepper noise at certain combinations of load and step rate, and the fact that we can't necessarily go from stationary to maximum achievable step speed instantly. At slow speeds and/or with light loads, the motor will snap to each new equilibrium point very quickly compared to the step rate, and will in fact stop at the bottom of the equilibrium valley waiting for the next change of magnetic state. This is *horrifically* wasteful of energy—every single step, we're injecting enough energy to overcome static friction forces and impart some amount of angular momentum to the stepper shaft and its attached load. The motor responds quickly and reaches its new stable state, but our control software isn't ready to move on to the next step state yet. The motor's stored inertia causes it to start climbing the slope of the next "hill," which turns the stepper temporarily into a generator. The energy we just pumped in is shorted through the transient protection circuitry and turned into heat. The motor slumps back down to the stable position until the next step pulse comes along, whereupon the entire process is repeated. This, along with the possibility of simple mechanical resonances in the motor and other components at certain step speeds, creates a loud and objectionable noise.

The most efficient way to drive the motor is to transit to the next step state as soon as the rotor reaches its new equilibrium position; this way, you are always pushing the load in the desired direction. Knowing exactly when this occurs is difficult, and it leads nicely into the next issue, which explains my point #2 on p. 66: We're not exerting a constant force on the rotor; rather, we are kicking it periodically. In order for our kicks to do the most good, we have to time them to coincide with certain rotor positions. At slow speeds, the motor responds to the step pulses faster than we issue them (so the motor can be said to be "led" by the step pulses). However, as the speed increases, we start to rely increasingly on the fact that the motor's inertia will carry it to the equilibrium point by the time we kick it next. This is why you can't simply start the motor at its maximum possible step speed—starting from rest, the rotor won't have time to reach equilibrium point before we transit to the next state.

On the flip side of the coin, once the motor is whirling madly at high speeds, simply stopping the step pulses dead will very likely lead to mechanical overshoot. You probably won't observe that phenomenon on a bare motor— at least, not the small, light motors you're likely to be using—but it's a very real issue under load. The most common approach to this is to implement an "acceleration profile," which is a table of step rates describing how to get from one motor speed to another most efficiently and reliably. These sorts of tables are easily handled in small microcontrollers. Advanced stepper designs will also use higher drive current (in other words, steeper hills) to achieve faster speed changes.

Completely general solutions to these problems (and others not mentioned) are possible—for example, I should point out that as soon as the motor overshoots the current equilibrium position, it will start generating a back EMF that can be measured via a suitably delicate circuit. However, such complex solutions are not required for simple, low-speed stepper applications, and they lie beyond the scope of this text.

3.5 Speed-Controlled DC Motor with Tach Feedback and Thermal Cutoff

E-2's main propulsion system consists of two DC motors directly driving contra-rotating propellers. An underwater vehicle with a single propeller is subject to undesirable torque forces, especially if the vehicle has no significant keel. It's possible to counteract this by using a stator to straighten the water flow behind the propeller. It's also obviously possible to drive two propellers from a single motor using gears. However, using two independently-controllable motors allows us to tighten the vehicle's turning circle by running the motors in opposite directions, if desired. It also lets the device limp home if one motor fails, or one propeller happens to foul something.

The textbook circuit configuration for controlling a reversible DC motor is the H-bridge, illustrated (representatively with bipolar transistors) in Figure 3-6:

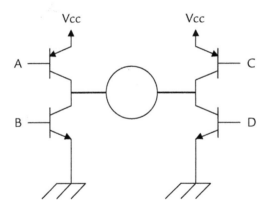

Figure 3-6: Standard H-bridge circuit

This is very much a "lowest common denominator" circuit, and although you might build one on a breadboard for a very quick and dirty test of something, you would never want to field a device built around such a simple configuration. However, it's a good starting point to illustrate the basic principles. To run the motor in one direction, turn on the PNP transistor on one side of the bridge (say, at A), and the NPN transistor on the other side (say, at D). To run the motor in the opposite direction, turn on the opposite pair of transistors; B and C in our example. You can control the motor's speed by modulating the on-time of either or both of the active transistors.

There are numerous practical problems with such a simplistic design. Perhaps most importantly, there is no protection for the switching transistors from the inductive "kick-back" from the motor windings. You could mitigate this by putting a protection diode across the collector and emitter of each transistor. Also consider what would happen if you reverse the motor direction by switching from the configuration (A ON – D ON) to the configuration (B ON – C ON). The switching times of the individual BJTs or FETs you would be using are not exactly identical, so you run the risk that, for an instant, both sides of the bridge could be "on," thus shorting the power rails—and probably either burning out part of the driver circuit, blowing a fuse or just causing a momentary power glitch that could reset some or all of your system to an unknown state. You could work around this problem by ensuring that the firmware never goes directly from the "powered up—forward rotation" to "powered up—backward rotation" states; instead, it should switch both sides of the bridge off for a brief recovery period before changing directions.

Furthermore, there is also no intrinsic hardware protection to prevent a firmware bug from shorting the power rails directly through one side of the bridge (for example, by switching on A and B simultaneously due to a software error writing garbage values to an I/O latch)—you could solve this by providing some external logic providing "direction" and "enable" inputs that only allow the drivers to be turned on in permissible combinations. Finally, as the circuit stands, you have no way to diagnose the health of the switching circuit or gauge the current being drawn by the motor, so you can't detect a stalled rotor or shorted winding.

Rather than reinventing all these wheels and engineering a custom solution, we cut around these messy problems by using the National Semiconductor LMD18200T integrated H-bridge. This chip is not exactly cheap, at around $11.50 (in single-piece quantities). However, the price is well worth the engineering time saved. If nothing else, you would probably spend at least twice this amount on destroying MOSFETs while debugging your own circuit design. The LMD18200 also offers several useful bonus features, including an internal junction temperature watchdog that will signal to your microcontroller with a simple digital signal if the chip is overheating (and shut the drivers off if the H-bridge overheats), integral shoot-through protection, nonregenerative braking (this shorts the motor windings) and a current monitoring output that, with an appropriate shunt resistor and ADC, can be used to measure how much current is being drawn by the motor. In fact, we won't be using the latter feature, so you might prefer to use the slightly cheaper LMD18201, which is identical to the LMD18200 except that it doesn't have the handy drive-current-monitoring feature. The reason I specify the LMD18200 is simply because it seems to be stocked by more vendors than its cheaper sibling. The price difference is only a few pennies from the distributors I use, but maybe you'll come across a load of amazingly cheap LMD18201s in the surplus marketplace.

There is one more feature our circuit offers, which isn't always essential but is often useful—tachometer feedback. Without some feedback on the actual physical number of revolutions being executed per second, it is practically impossible to control the speed of a motor under varying load. The tach input of the board expects to see an active low pulse once per revolution. The sensor method I use on the E-2 is a Hall effect sensor mounted next to the motor shaft, and a tiny neodymium magnet glued to the shaft. You might prefer to use some other method, such as an optical sensor and a reflective (or dark) mark painted on the shaft. Some motors even have tach hardware of some sort built in; this is particularly common in small cooling fans, which frequently have an integral Hall effect sensor. With such motors, all you need to do is connect the wires properly, and you're done[16]. However, you should note that the tach on these motors may not be reliable at anything less than 100% PWM duty cycle. These cooling fans are often designed to run continuously at full speed, with the tach providing feedback to the system that the fan hasn't stalled. The tach sensor is probably powered directly off the power input wire, and may not have enough of a decoupling capacitor connected to remain alive during the "off" portion of your PWM signal.

Enough talk, let's look at the schematic for our motor controller:

---

[16] Note: Almost all fans output two tach pulses per revolution. Depending on what kind of tach sensor you employ, and how you mount it, your system may output only one pulse per revolution.

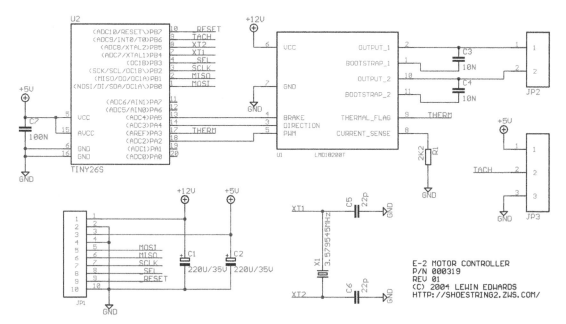

Figure 3-7: DC motor control circuit

The first thing you'll notice is that it's quite difficult for us to use the tiny26L's inbuilt PWM hardware, because its functions overlay the pins used by the USI. If you study the tiny26L's datasheet closely, you'll observe that we *could* use PB3 (which doubles as the OC1B output) as the PWM drive signal. However, this would mean either moving the SPI select functionality to a noninterrupt pin—unacceptable, because we have to respond to changes in the select state quickly so we can get off the bus—or enabling the second pin-change interrupt, PCINT1. The latter interrupt is a bit too "global" for comfort—it fires even when alternate hardware functions have been selected for the I/O pins. Extra care would be necessary to suppress these unwanted interrupts.

As a result, we have to use a software PWM drive scheme. Essentially, I had to choose between either software USI or software PWM, and in order to keep as much code as possible common between the various modules of this project, I chose to do the PWM in software. An unavoidable side-effect of this is that serial I/O and tach pulses will cause minor glitches in the PWM output. A second side-effect is that the PWM frequency is severely limited by the AVR's core clock speed. (The hardware

73

PWM feature can be driven by an asynchronous clock generated by an on-chip PLL). Neither of these are serious problems for our application.

Observe also that we use the dedicated INT0 line for the tachometer input. The reason for this is that we're only interested in one edge of the tach signal. Although we've got enough free GPIOs to dedicate one of the pin-change interrupts to tach input, this interrupt would fire on both edges of the tach signal, which introduces un-necessary glitches into our other tasks. INT0 can be programmed to fire only on the edge of our choice.

Now let's study the firmware for this device. It's moderately complicated, because there is a lot going on, so let's study the major tasks separately. We'll first look at the task that handles the PWM:

```
;====================================================================
; Timer 0 overflow - PWM motor output
; WARNING - This code must be bare-minimum optimized. It runs so
; frequently that it can easily take over the entire CPU.
entry_timer0:
        in r3, SREG
        sbi PORTA, PORTA_TEST

        ; Reset TMR0 counter to start position for next tick
        ldi r24, TMR0_RELOAD
        out TCNT0, r24

        ; Increment PWM count variable
        inc r22
        cpi r22, 100
        brne pwm_no_overflow

        ; If we've overflowed, reset to 0. Do a special-case check to see
        ; if the requested duty cycle is 0 - if so, turn off the PWM out-
put.
        ldi r22,0
        cp r1, r22
        brne pwm_duty_nonzero

        cbi PORTA, PORTA_PWM
        rjmp tmr0_done
```

```
pwm_duty_nonzero:
        sbi PORTA, PORTA_PWM
        rjmp tmr0_done

        ; If we didn't overflow the 100% counter, then compare to
        ; requested duty cycle.
        ; If count = duty cycle, turn off the PWM output. Note that the
        ; 100% special case is already handled by the overflow trapping
        ; above.

pwm_no_overflow:
        cp r22, r1
        brne tmr0_done
        cbi PORTA, PORTA_PWM

tmr0_done:
        cbi PORTA, PORTA_TEST
        out SREG, r3
        reti
```

There are a few parameters that need to be balanced here:

- The desired number of PWM "grayscales" between 0% (always off) and 100% (always on) duty cycle. We're going to use 100 levels, so the PWM value in use is actually the same number as the duty cycle percentage. There are a couple of reasons for choosing this value besides the mere elegance of specifying a percentage directly: firstly, since the AVR is an 8-bit micro, it's convenient for us to use a value that fits in an 8-bit variable with some wiggle room so we can avoid fencepost errors[17] without having to insert any unwieldy special-case code for boundary conditions. Secondly, 100 steps is really more than enough for most drive motor applications; in fact, E-2 only adjusts the duty cycle in 10% increments.

---

[17] Imagine you are building a fence 100 meters long. The posts along this fence are ten meters apart. How many posts do you need to buy? The instinctive response of ten is the canonical fencepost error.

■ The target PWM frequency. 200 Hz is a workable (though noisy) frequency for motor driving. To give an admittedly oversimplified rule of thumb: ultrasonic frequencies are generally better for PWM applications, because they push the pulse noise out beyond our hearing range. However, the tiny26L is too slow to implement this in software. If we were using the hardware PWM, it wouldn't be a problem, but unfortunately we don't have that luxury, for reasons described previously. 200 Hz is a reasonable compromise.

■ System clock speed and interrupt loading issues.

PWM in our design is implemented using Timer 0. This 8-bit timer increments with a selectable ratio (/1, /8, /64, /256 or /1024) of the system clock frequency, and interrupts when it rolls over from 0xFF to 0x00. The parameters we can vary here are the clock divisor and the value to be reloaded into the timer during the interrupt. In order to calculate these values, we need to work backwards from the 200 Hz figure. In each cycle of that 200 Hz signal, we need to have 100 sample points at which the PWM drive signal can be either off or on. This means we need a timer interrupt rate of 20 kHz. Given our system clock of 3.579545 MHz, the Timer 0 clock frequencies available to us are approximately 3.58 MHz, 447 kHz, 55.9 kHz, 14.0 kHz or 3.50 kHz. The slowest timer speed that is faster than our 25 kHz target is 55.9 kHz. Dividing these frequencies we see that we need to divide the timer 0 clock source (by reloading the timer with some nonzero value) by 2.24 to get the right interrupt rate. The closest we can get is either a reload value of 0xFE (2 ticks per interrupt, 28.0 kHz) or 0xFD (3 ticks per interrupt, 18.6 kHz). We'll use 0xFE, because even though it represents a greater error with respect to our nominal target frequency, it's an error "in the right direction"—that is, towards better performance.

As an aside at this point, you'll notice by inspecting the subroutine shown on the previous page that my code sets PA7 high on entry to the Timer 0 interrupt, and brings it low again just before returning from that ISR. The purpose of this strobe is so that we can observe on an oscilloscope the amount of CPU time being chewed up by the PWM function. Since this is the most frequently-executed code path in the chip, it's instructive to have some means of measuring how big a timeslice it occupies. From quick inspection, rather less than 10% of the available CPU time is being occupied in the PWM ISR, which is quite acceptable. While we're talking about performance, though, also note how I've fine-tuned the register usage in this project to avoid having to save anything on the stack in the high-load ISRs. It gets harder and harder to do these kinds of down-to-the-last-nibble optimizations as you add more tasks to a system (simply because each task has a certain amount of state that needs to be stored). This is another useful argument in favor of breaking up a system's real-time responsibilities across multiple microcontrollers.

The second task in this module is tachometer measurement. This is handled by Timer 1. Timer 1 is a free-running up counter. Timer 1 has a richer selection of prescaler values than Timer 0, from /1 (3.58 MHz) all the way down to /16384 (218 Hz). What scaler value should we choose? It depends on the desired accuracy and anticipated range of the input signal. Let's say that we are going to use a motor with a maximum unloaded speed of 3,600 rpm and arbitrarily pick a divisor of CK/8192, or 437 Hz. To get a reasonably accurate measurement, we are actually going to perform a kind of running average: The tach interrupt increments a 16-bit counter, but does NOT permit it to roll over past 0xFFFF. Timer 2's overflow interrupt is allowed to fire eight times (2048 ticks, or in other words ~4.69 seconds). At the eighth interrupt, the tach counter is captured and reset; it can range from 0 (no motion) to 0xFFFE (838,000 rpm; exceedingly unlikely). For an unloaded motor of the type you're likely to be using, expect to see values around 274, (3,500 rpm). 0xFFFF values should be displayed by your user interface as "out of range," and 0 should probably be displayed as "tach failure," since these are extreme boundary conditions. Note that you can tinker with the dynamic range and sample rate of this measurement very simply by altering the tachometer divisor value in the Timer 1 interrupt. The actual code for the two tachometer-related ISRs are as follows:

```
;========================================================================
; Timer 1 overflow - Tach sampler
entry_timer1:
        in r5, SREG

        inc r21
        cpi r21, TACH_DIVISOR
        brne tmr1_done

        ldi r21, $00
```

```
        ; Update RAM copy of counter
        sts tach_low, r30
        sts tach_high, r31

        ; Clear tach counter
        ldi r30, $00
        ldi r31, $00

tmr1_done:
        out SREG, r5
        reti


;========================================================================
; Tachometer input handler (INT0)
; This interrupt fires once every revolution, and would typically
; be triggered by a stationary Hall effect sensor sensing a magnet on
; the shaft.
entry_int0:
    in r4, SREG

        ; Increment tach counter
        inc r30
        brne tach_done

        inc r31
        brne tach_done

        ; If tach has overflowed, peg it at $FFFF.
        ldi r30, $ff
        ldi r31, $ff

tach_done:
        out SREG, r4
        reti
```

Since this module isn't directly concerned with the actual motor speed, we just provide the above 16-bit counter result when queried for status—the host is expected to do the math to convert the raw count into a rotation speed in a unit acceptable to the end-user. The value can be calculated very simply by:

$$\text{speed in rpm} = (\text{tach value} * 60) / 4.69.$$

The final task is the SPI interface management code. This code is very similar to the analogous portions of the stepper motor controller; a simple state machine which is stimulated by either the USI interrupt (for incoming data) or the pin-change interrupt that handles SPI slave selection and bus-on/-off events. The firmware supports the following command codes:

Mnemonic	Value	Description
MTR_CMD_STOP	0	Stop motor.
MTR_CMD_FWD	1	Set PWM duty cycle and start motor spinning in the forward direction. The next byte following this command byte should be the duty cycle (0–100%).
MTR_CMD_REV	2	Set PWM duty cycle and start motor spinning in the reverse direction. The next byte following this command byte should be the duty cycle (0–100%).
MTR_CMD_READ_STATUS	254	Reads back current speed and tachometer status. The host can read back up to three bytes following this command; the first byte is the status/speed byte (lower 7 bits = current PWM duty cycle, upper bit is set if the thermal warning flag is active), and the next two bytes are the most recent tach count, low byte first.
MTR_CMD_RESET	255	Performs a soft reset.

3.6 Two-Axis Attitude Sensor using MEMS Accelerometer

For a variety of reasons; navigation, hazard avoidance, and so on, it's desirable for a vehicle to be able to know its orientation with respect to the earth. A ship, submarine or airplane has six degrees of freedom (land-bound vehicles generally have fewer). Three of these are rotational: rotation around an imaginary line from bow to stern (roll), rotation around an imaginary line perpendicular to the bow-stern line and parallel to the Earth's surface (pitch), and rotation in a plane parallel to the Earth's surface (yaw; turning the bow of the vehicle to point towards a new destination). The other three are translational, along the same axes just mentioned;

respectively, surge (movement forwards or backwards), sway (movement from side to side) and heave (movement up or down).

Yaw is relatively difficult to measure directly, so let's discuss it first. One approach is to use a flux-gate sensor; an electronic compass, essentially. The difficulty with this is that every spot on the Earth's surface has more or less interference from local metallic deposits and other geographical features, so a compass needle doesn't always point at a known reference point (magnetic north). Magnetic north also moves about, and it doesn't coincide with the true geographic north pole. For short trips, a fixed variance setting can be looked up on a map of your area, and you can just ignore any errors caused by roaming about close to vast lodestone deposits! E-2's core electronic module doesn't directly measure yaw; it assumes that most of the vessel's motion vector is parallel to the bow-to-stern axis and hence uses GPS velocity data (while surfaced) to infer the direction the bow is pointing. If you want to try your hand at magnetic navigation methods, there are numerous kits containing flux-gate compass boards, intended for the hobbyist robotics market. Most of these incorporate some clever firmware to deal with variance issues.

Static roll and pitch, on the other hand, can easily be ascertained by measuring the gravity vector acting on the craft and comparing it to an imaginary reference vector at right angles to both the bow-stern and port-starboard axes of the vehicle. To perform this measurement task, we use an accelerometer.

At its simplest, an attitude or acceleration sensor is simply a pendulum. In fact, a reasonably useful two-dimensional attitude sensor can be constructed by simply taking a two-axis potentiometer assembly out of an off-the-shelf analog joystick, attaching a heavy weight to the joystick lever, and mounting the whole thing upside-down so that the weighted joystick can swing around freely. (Pay attention to align the axes of the sensor with the axes around which the sensor is expected to rotate). In cases where fine accuracy is not essential and it is desirable to connect this sensor directly to a PC, cannibalizing a joystick in this way is definitely the path of least resistance, not to mention an extremely fast way to construct a prototype. Apropos of the E-2 project, it is interesting to note that attitude and depth control in torpedoes of World War II vintage were actually controlled using a mechanically interlinked system of a pendulum and a manometer.

Despite its simplicity, there are a number of disadvantages to the simple pendulum method—it is bulky, and friction in the potentiometers and joystick bearings tends to makes the device insensitive to small accelerations. A better solution for some applications is to use free-turning weights or gyroscopes on exquisitely low-friction bearings, with some sort of optical or magnetic scale read-out, but these sorts of machines are expensive and relatively high-maintenance. The modern solution to this design problem is an integrated MEMS (MicroElectroMechanical System) part such as the Analog Devices' ADXL202 two-axis accelerometer. This particular device is only rated up to two gravities (approximately 19.7 ms^{-2}) of acceleration. This makes it suitable for assessing the overall attitude of a body, but not terribly useful for more demanding tasks. For truly challenging tasks like measuring deceleration during a car crash, or rocket takeoff forces, you need (at the very least) a part rated for much higher accelerations.

Before we go any further, please note that this text talks specifically about the older ADXL202JQC (commercial temperature grade) and ADXL202AQC (industrial temperature grade) parts, which were available in a 14-lead ceramic SOIC package. This variant has been discontinued by Analog Devices in favor of a ceramic 8-pin leadless chip carrier, the ADXL202JE (commercial) and ADXL202AE (industrial). However, the older part is still available in distribution channels and is quite widely used in hobbyist type applications because of the relative ease with which it can be hand-prototyped.

The best way to prototype with either part is, of course, with a small PCB or the evaluation board for the part. If this is not possible (note that the EVB is approximately three times the cost of the bare part in single-piece quantities), then an acceptable alternative for the older SOIC part is to glue the sensor to the non-coppered side of a piece of protoboard, and solder thin wires to the pins (wire-wrap type wire performs this duty very well). I prototyped the circuit in this book using this method; I used a two-part epoxy resin to glue the chip down. If you are using the more modern LCC part, however, life is more difficult. Here's a picture of the older device (glued to a piece of prototype board), alongside a couple of samples of the LCC part, one of which has been turned upside-down to show the contact pattern on the underside. Those contacts are very fine gold deposits on the ceramic chip body; if you solder wires to them, you can quite easily pull the contacts right off the chip.

Nevertheless, if that's your only prototyping option, it *can* be done—just be very careful not to apply any unnecessary stress, because those parts are expensive (about $15–20 each, in small quantities).

Figure 3-8: Different ADXL202 variants.

Mechanical package considerations of this sort aside, the ADXL202 is very easy to interface. The device outputs two square wave signals (one for each axis) with an identical period, T_2, and a variable duty cycle with an on-time of T_1. Note that although the two signals are guaranteed to have the same period, they are *not* guaranteed to start at the same time—they can have any amount of phase difference. The period can be configured from 1 to 10 ms by means of an external resistor (R_{SET} in Analog Devices literature), selected according to the formula:

$$T_2 = R_{SET}/125000000,$$

where T_2 is in seconds, and R_{SET} is in ohms. I have chosen a 1M resistor, which gives us a nominal period of 8ms. Actual measurement of a real device in-circuit shows a period of 7.2ms, which is gratifyingly close to the mark. Rather than sketch the waveforms artificially, here is a picture of an actual scope trace showing both X (bottom) and Y (top) outputs for an ADXL202JQC. The device generating these signals was flat on my desk, which is approximately horizontal with reference to the Earth's surface.

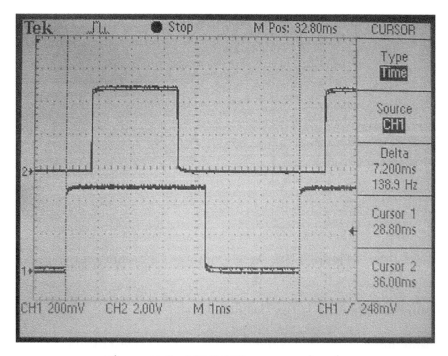

Figure 3-9: ADXL202 output signals

The on-time period T_1 (the "hump" in the waveforms in Figure 3-9) is nominally supposed to be 0.5 T_2 when the acceleration on the axis in question is 0g. In practice, though, there is a wide deviation—as you can see from the measurements in Figure 3-9, where the accelerometer was known to be approximately horizontal.

Following is the schematic for our circuit:

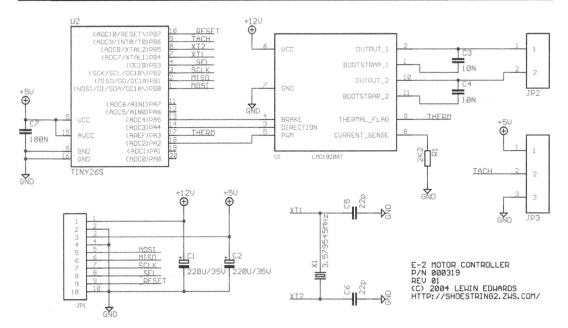

Figure 3-10: Accelerometer schematic

The firmware supports only two command codes:

Mnemonic	Value	Description
ACL_CMD_READ_STATUS	254	Reads back six bytes of accelerometer status.
ACL_CMD_RESET	255	Performs a soft reset.

ACL_CMD_READ_STATUS returns three 16-bit words of status information; first, the measured value of T_2, then the T_1 value measured for the X axis, and finally the T_1 value measured for the Y axis. The high-order byte is transmitted first. Observe that we don't need to calibrate and transmit two copies of T_2, since it is known (by design) to be identical for both axes.

The meat of this project is contained in the Timer 1 and pin-change interrupt handlers. The interrupt handler for overflows in Timer 1, which runs at the full CPU clock speed, merely increments a high-order counter byte, thereby extending Timer 1's range to 16 bits. This corresponds to a maximum measurable T_2 of approximately 18ms, with a theoretical resolution of 279 ns.

```
;========================================================================
; Timer 1 interrupt
entry_tmr1:
      in r4, SREG

      inc r1

      out SREG, r4
      reti


;========================================================================
; I/O pin change interrupt
; The only valid sources of this interrupt are PB3, which is used as
; the 3-wire slave select line, PA6 (X-input) and PA7 (Y-input)
entry_iopins:
   in r3, SREG

      ; Grab current timer value in case we need it later
      in r22, TCNT1

      ; Get last value of shadow register
      lds r26, porta_shadow

      ; Scan X-input. First, handle the case if it is high.
      sbis PINA, PORTA_X
      rjmp x_is_low

      sbrc r26, PORTA_X
      rjmp test_y               ; No change. Test Y-input.

      ori r26, PORTA_X

      ; +ve edge detected on X. We need to calculate T2
      ; by subtracting last-edge from current timer, and adding X-T1
      ; to that result.
      mov r21, r22              ; timer lo byte
      lds r25, xle_lo
      sub r21, r25              ; r21 = intermediate val lo byte
      mov r24, r1               ; timer hi byte
      lds r25, xle_hi
      sbc r24, r25              ; r24 = intermediate val hi byte
      lds r25, x_t1_lo
      add r21, r25
```

```
        lds r25, x_t1_hi
        adc r24, r25
        sts t2_hi, r24
        sts t2_lo, r21

        rjmp update_x_edge

        ; X-input is low. Test for -ve edge.

x_is_low:
        sbrs r26, PORTA_X
        rjmp test_y

        ; -ve edge on X detected. We need to calculate X-T1
        ; by subtracting last-edge from current timer.
        mov r21, r22            ; timer lo byte
        lds r25, xle_lo
        sub r21, r25            ; r21 = T1 lo byte
        mov r24, r1                    ; timer hi byte
        lds r25, xle_hi
        sbc r24, r25            ; r24 = T1 hi byte
        sts x_t1_hi, r24
        sts x_t1_lo, r21

        ; Update last-edge timestamp for X axis

update_x_edge:
        sts xle_hi, r1
        sts xle_lo, r22

        ; Scan Y-input - First, handle the case if it is high.

test_y:
        sbis PINA, PORTA_Y
        rjmp y_is_low

        sbrc r26, PORTA_Y
        rjmp test_spi           ; No change. Go to SPI test.

        ; +ve edge detected on Y. We need to start calculating Y-T1.
        rjmp update_y_edge

y_is_low:
        sbrs r26, PORTA_Y
        rjmp test_spi           ; No change. Go to SPI test.
```

```
        ; -ve edge on Y detected. We need to calculate Y-T1
        ; by subtracting last-edge from current timer.
        mov r21, r22          ; timer lo byte
        lds r25, yle_lo
        sub r21, r25          ; r21 = T1 lo byte
        mov r24, r1                   ; timer hi byte
        lds r25, yle_hi
        sbc r24, r25          ; r24 = T1 hi byte
        sts y_t1_hi, r24
        sts y_t1_lo, r21


        ; Update last-edge timestamp for Y axis

update_y_edge:
        sts yle_hi, r1
        sts yle_lo, r22

test_spi:
        in r26, PINA
        sts porta_shadow, r26

        ; Check state of SPI select line.
        sbic PINB, PORTB_SEL
        rjmp usi_disable

        ; If PB1 is already an output, don't reset the USI. This
        ; special code is necessary so accelerometer interrupts don't
        ; mess with partially complete USI transactions.
        sbic DDRB, PORTB_DO
        rjmp iopin_done

        ; SEL line is LOW. Enable and reset USI and switch PB1 to output
        ldi r23, FSMS_RXCMD
        ldi r26, $00
        out USIDR, r26        ; Empty USI data register
        out USISR, r26        ; Clear USI status (including clock count!)
        sbi DDRB, PORTB_DO    ; set PB1 to output

        sbi USISR, USISIF     ; Clear start condition status
        sbi USISR, USIOIF     ; Clear overflow status
        sbi USICR, USIOIE     ; Enable USI overflow interrupts

        rjmp iopin_done
```

```
        ; SEL line is HIGH. Disable USI and switch PB1 to input to take
        ; us off-bus

usi_disable:
    ; disable USI start and overflow interrupts
        cbi USICR, USISIE
        cbi USICR, USIOIE

        ; Disable output driver on PB1 (DO)
        cbi DDRB, PORTB_DO        ; set PB1 to input

iopin_done:
        out SREG, r3
        reti
```

Some averaging or filtering is advisable on the PC end of this equation. Here's a simple program that takes out continuous readings and prints them to a single line on the console (you'll find the sourcecode and makefile for this program in the accel directory of the sample source archive):

```c
/*
    main.c

    Demonstration applet for E2BUS stepper interface code

    From "Open-Source Robotics and Process Control Cookbook"
    Lewin A.R.W. Edwards (sysadm@zws.com)
*/

#include <stdio.h>

#include "e2bus.h"

int main (int _argc, char *_argv[])
{
    unsigned char pkt[6];
    int i=0,j;

    // Open port
    if (E2B_Acquire()) {
        printf("Error opening E2BUS.\n");
        return -1;
    }
```

```
E2B_Reset();
printf("Reset complete, pausing...\n");
sleep(1);
printf("XXXXXXXXXXXXXXXXXXXXXXXXXXXXXXXXXXXXX");
while (1)
{
   usleep(750000);
   pkt[0]=ACL_CMD_READ_STATUS;
   E2B_Tx_Bytes(pkt, 1, 0, 0);
   E2B_Rx_Bytes(pkt, 6, 0, 1);
   for (j=0;j<37;j++) printf("\b");
   printf("Sample %-08.8X status %-02.2X%-02.2X,"
      "%-02.2X%-02.2X,%-02.2X%-02.2X",
      i, pkt[0],pkt[1],pkt[2],pkt[3],pkt[4],pkt[5]);
   fflush(stdout);
   i++;
}

   return 0;
}
```

If you run this program, you'll see that even if the accelerometer is stationary, there's a certain amount of jitter in the output values. This is partly due to the irritatingly analog nature of the Universe, partly due to vibration of the accelerometer, and partly due to the fact that serial interrupts can slightly skew the time measurement task. For example, here is a sequence of three consecutive readings from my prototype:

6821, 2C0A, 3AD6

67D7, 2BFC, 3AF0

67E9, 2C46, 3A2C

Because this phenomenon is unavoidable, some averaging or filtering is desirable before working with the sensor output. Simple averaging is acceptable, but a Kalman filter is better; for more information on this topic, I recommend the reference **du Plessis, R.M.**, 1967; *Poor man's explanation of Kalman Filters or How I stopped worrying and learned to love matrix inversion*, ISBN 0-9661016-0-X.

3.7 RS-422—Compatible Indicator Panel

This circuit is a bit of a departure from the rest of the content in this chapter, inasmuch as the appliance I describe here is not part of the E-2 project. The reason I have included this section is because the circuit and firmware illustrate several relevant and interesting points, including multidrop differential serial communications over relatively long distances and using the AVR's internal RC oscillator instead of an external crystal. This application also provides a nice example of how the sorts of systems in this book can be used in real-world situations.

I developed this device for a shipping center application used in two of a company's warehouses. The products shipped from these centers consist of standard and customized kits of individually-packaged parts. A number of conveyor belts run through the warehouse area, past the various bins of parts. At the "start" end of each conveyor belt is a large matrix of 416 pigeonholes arranged as 26 rows by 16 columns. Before each shift, administrative staff stock these pigeonholes with picklists describing different standard subassemblies. Each pigeonhole has an indicator lamp (actually, an LED) over it. A central computer, connected to the company's order processing system, controls all these indicator lamps over a piece of Category 5e cable that runs approximately 800 feet from the computer room to the warehouse floor; the indicator panels show workers along the conveyor belt which pick-lists to gather for an individual order as it progresses down the line. As initially installed, all the panels were to repeat a single set of commands, however it was desired to leave the functionality open-ended so that in future, more panels could be added to the same bus, but show a different set of signals (to process multiple orders simultaneously on the same line). For this reason, each panel has an 8-bit address; commands coming down the wire have an address field indicating the intended recipient. It's legal for multiple indicators to have the same address if you want them to repeat duplicate data.

Following is the schematic for our circuit:

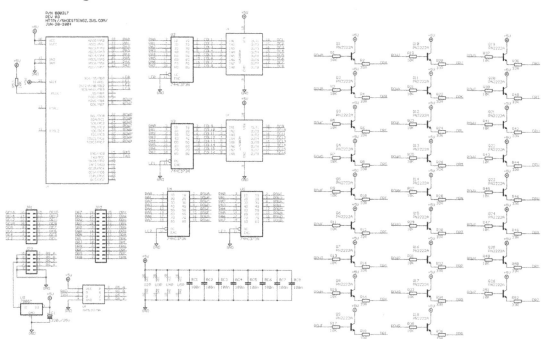

Figure 3-11: Schematic for RS-422-compatible indicator panel

The actual LEDs are omitted from this schematic for clarity's sake; they are wired in a simple matrix with the cathodes connected to the ULN2803s driving the column lines, and the anodes connected to the row lines.

I chose an ATmega16 part for this design purely for the large I/O and memory budget; .although it would be possible to implement the project in a much smaller part, it was simply convenient and quick to pick the mega16. You'll observe that this project uses the AVR's on-chip clock generator rather than relying on an external crystal. Note that the internal RC oscillator in the AVR parts is factory-calibrated with a device-specific "fudge factor." This fudge factor is different for each supported oscillator speed. The specific calibration constants for each frequency are stored in a nonuser-accessible (probably OTP) area of the micro, and can be read out with the chip signature. They cannot be read directly by code running on the target device;

you can only read them out with a device programmer like the STK500. You'll notice that the code I provide is set up for 1 MHz operation, but that I have also included commented-out initialization code for 8 MHz operation. If you want to run the project at 8 MHz, you need to do slightly more than just uncomment the faster initialization code, though—you need to make sure the chip is correctly initialized with the right oscillator fudge factor for 8 MHz. It's something of a design shortcoming in the AVR series, but when the device is configured for any RC oscillator mode, it automatically loads the 1 MHz calibration factor (the first calibration byte) into the processor's oscillator calibration register, *regardless of what RC oscillator speed you have selected*. If uncorrected, this can lead to considerable clock deviation from the expected speed. The oscillator might not even be reliable if it's grossly miscalibrated.

Atmel's official suggestion is to use an EEPROM or program flash location to store the factory calibration value, and copy it into the oscillator calibration register at power-up. If you run AVR Studio, select Tools—STK500/AVRISP/JTAG ICE —STK500/AVRISP/JTAG ICE, and click the Advanced tab, you'll be able to read the signature bytes out of the chip using the Read Signature button. Select the desired speed from the drop-down list in the Oscillation Calibration byte section, and click "Read Cal. Byte." The appropriate calibration value will appear in the Value: box. (Remember—this value is specific to the individual chip you're looking at; you can't reuse this calibration value in a different chip). You can now select a flash or EEPROM address using the fields under Write Address, and click "Write to Memory" to copy the calibration byte into flash or EEPROM, as illustrated by the following two screenshots:

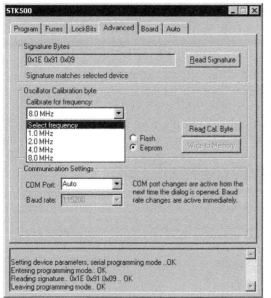

Figure 3-12:
Select the desired AVR clock speed

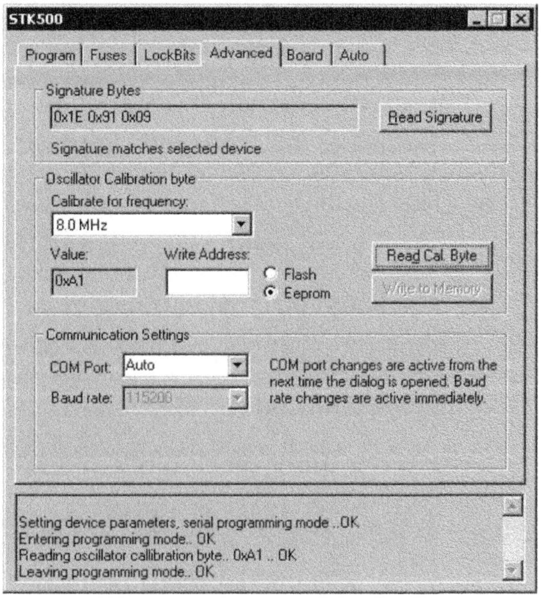

Figure 3-13:
Read the appropriate calibration byte

While we're on the topic of burning chips, note that the ATmega16 ships with JTAG enabled by default. You need to set the fuses to disable JTAG in order to free up the associated I/O pins for use as general-purpose I/O in our application.

The indicator panel's serial interface operates at 2400 bps with 8 data bits, no parity and one stop bit. This panel is operated using a series of command strings, formatted as follows (note that these strings are case-sensitive):

- An attention character, '!'.

- An address character indicating which board(s) should hear the message. The code in this book has been hardcoded to use address 'A' (65). Refer to the sourcecode to change the unit address. If multiple indicator panels on the same bus have the same address, they will all respond to messages at the same time.

- A command character, which is one of: 'R' (reset; turn off all LEDs), '1' (turn on specified LED), '0' (turn off specified LED), 'B' (turn on blinking for a specified column), 'b' (turn off blinking for a specified column), or 'T' (start test mode; test mode runs until canceled with the 'R' command).

- The 1 and 0 commands require two additional bytes—a column identifier (A~P, corresponding to columns 0 through 15) and a row identifier (A~Z).

- The B and b commands require one additional byte identifying the column to blink (A~P, corresponding to columns 0 through 15).

Example command strings:

!AR   - Turns off all LEDs on unit with address A.

!A1BG   - Turns on the LED (on unit with address A) at column 1, row G.

!AT   - Starts test mode on unit with address A.

!A0FK   - Turns off the LED (on unit with address A) at column 5, row K.

!ABD   - Starts blink mode for column 3 (on unit with address A).

!AbD   - Stops blink mode for column 3 (on unit with address A).

Since this is the most complex project, by code volume, included with this book, I am detailing the entire sourcecode in the text.

```
;=======================================================================
; Miscellaneous constants
.equ   MY_ID           =65        ; ID number of this unit (default 'A')
.equ   BLINK_RATE      =30        ; Frames per blink-toggle

;=======================================================================
; Special serial Rx characters
.equ   CHR_ATTENTION   =$21            ; !
.equ   CHR_RESET       =$52            ; R
.equ   CHR_TESTMODE    =$54            ; T
.equ   CHR_LEDOFF      =$30            ; 0
.equ   CHR_LEDON       =$31            ; 1
.equ   CHR_BLKON       =$42            ; B
.equ   CHR_BLKOFF      =$62            ; b
```

```
;=====================================================================
; States for serial Rx state machine
.equ   SRX_WAIT        =0          ; Wait for attention character
.equ   SRX_ID          =1          ; Wait for unit ID
.equ   SRX_GETCMD      =2          ; Wait for command byte
.equ   SRX_ON_GETCOL   =3          ; Wait for column byte for LED-on command
.equ   SRX_ON_GETROW   =4          ; Wait for row byte for LED-on command
.equ   SRX_BLKON_COL   =5          ; Wait for column for blink-on command
.equ   SRX_BLKOFF_COL  =6          ; Wait for column for blink-off command
.equ   SRX_OFF_GETCOL  =7          ; Wait for column byte for LED-off command
.equ   SRX_OFF_GETROW  =8          ; Wait for row byte for LED-off command


;=====================================================================
; Bits in flags
.equ   FLAG_BLINK      =7          ; Blink flag, toggled every BLINK_RATE
frames
.equ   FLAG_TESTMODE   =6          ; Nonzero = unit in test mode



;=====================================================================
; Variables in SRAM
.DSEG
.ORG 0x60
currentline:    .BYTE 1           ; Column# currently being driven
frameptr_lo:    .BYTE 1           ; Pointer to frame data for current column
frameptr_hi:    .BYTE 1
framecounter:   .BYTE 1           ; Incremented each frame refresh
flags:          .BYTE 1
serialmode:     .BYTE 1           ; serial FSM code
tmpcol:         .BYTE 1           ; Temporary holding buffer for column#

.ORG 0x80
; BUGBUG Do not move this structure. The arithmetic that works with
; it will break if this 64-byte structure crosses a 256-byte boundary.
; For safety, leave it here.
; Each table entry is formatted as follows:
; BYTE   -   A-H
; BYTE   -   I-P
; BYTE   -   Q-X
; BYTE   -   Z (bit 7), Y (bit 6), blink (bit 0) - other bits reserved,
leave 0
framedata:      .BYTE (4*16)    ; Each 4 bytes is a column of LED data
```

```
;=====================================================================
; Interrupt vectors
; This must be the first thing in the executable flash image.
.CSEG
.ORG 0x0000
      jmp       entry_reset      ;Reset
      jmp       bad_irq          ;External interrupt request 0
      jmp       bad_irq          ;External interrupt request 1
      jmp       bad_irq          ;Timer/Counter2 Compare Match
      jmp       bad_irq          ;Timer/Counter2 Overflow
      jmp       bad_irq          ;Timer/Counter1 Capture Event
      jmp       bad_irq          ;Timer/Counter1 Compare Match A
      jmp       bad_irq          ;Timer/Counter1 Compare Match B
      jmp       bad_irq          ;Timer/Counter1 Overflow
      jmp       tc0_overflow     ;Timer/Counter0 Overflow
      jmp       bad_irq          ;Serial transfer complete
      jmp       usart_rx         ;USART Rx Complete
      jmp       bad_irq          ;USART data register empty
      jmp       bad_irq          ;USART Tx Complete
      jmp       bad_irq          ;ADC conversion complete
      jmp       bad_irq          ;EEPROM ready
      jmp       bad_irq          ;Analog comparator
      jmp       bad_irq          ;Two-wire serial interface
      jmp       bad_irq          ;External interrupt request 2
      jmp       bad_irq          ;Timer/Counter0 Compare Match
      jmp       bad_irq          ;Store Program Memory Ready


;=====================================================================
; Invalid exception handler
bad_irq: ldi r16, $BB
         out PORTA, r16
         rjmp bad_irq


;=====================================================================
; Power-on reset entry point
entry_reset:
      ; Initialize stack pointer to top of RAM
      ldi r16, high(RAMEND)
      out SPL, r16
      ldi r16, low(RAMEND)
      out SPH, r16

      ; Configure ports A,B,C as outputs
      ldi r16, $FF
```

```
        out DDRA, r16
        out DDRB, r16
        out DDRC, r16

        call clear_outputs

        ; clear status flags and serial FSM
        ldi r16, $00
        sts flags, r16
        sts serialmode, r16

        ; Configure PD1 (TxD) as output
        ldi r16,$02
        out DDRD, r16


;8MHz setup code
;       ldi r16, $CF
;       out UBRRL, r16
;       ldi r16, $00
;       out UBRRH, r16

        ; 1MHz clock setup
        ; Set up USART for 2400bps asynchronous mode
        ; Formula for calculating UBRR in this case is fosc/(16 * baud) - 1
        ldi r16, $19
        out UBRRL, r16
        ldi r16, $00
        out UBRRH, r16

        ; clear USART status
        out UCSRA, r16

        ; configure control/status register B
        ldi r16, (1 << RXCIE) + (1 << RXEN) + (1 << TXEN)
        out UCSRB, r16

        ; configure control/status register C
        ; 8 bits, no parity,
        ldi r16, (1 << URSEL) + (1 << UCSZ1)  + (1 << UCSZ0)

        ; set frame pointer to 0 and clear frame counter
        ldi r16, $00
        sts currentline, r16
        sts framecounter, r16
```

```
            ; Point X at start of framebuffer data
            ldi r26, low(framedata)
            ldi r27, high(framedata)
            sts frameptr_lo, r26
            sts frameptr_hi, r27

            ; clear timer counter
            ldi r16, $00
            out TCNT0, r16

            ; clear timer 0 interrupt flag
            ldi r16, $00
            out TIFR, r16

;  8MHz setup code
;           ldi r16, $03                      ; /64 prescaler
;           out TCCR0, r16

            ; 1MHz setup code
            ; set up timer 0 for fosc/8 (=125kHz)
            ; This corresponds to a line rate of ~977Hz, frame rate ~61Hz.
            ldi r16, $02                      ; /8 prescaler
            out TCCR0, r16

            ; enable timer 0 interrupt
            ldi r16, (1 << TOIE0)
            out TIMSK, r16

            ; enable interrupts
            sei

            call clearscreen
```

The USI-complete interrupt handles all the video write tasks:

```
;=======================================================================
; ISR
; Serial Rx complete
usart_rx:
            push r27
            push r26
            push r18
            push r17
            push r16
```

```
        push r0
        in r0, SREG

        ; Get received byte from USART
        in r16, UDR

        ; Act on the byte depending on the FSM state
        lds r17, serialmode

        cpi r17, SRX_WAIT
        brne rx_notwait

        ;================================================================
        ;Check for attention character
        cpi r16, CHR_ATTENTION
        breq rx_atn
        rjmp rx_done
rx_atn:

        ; Attention char received! Now wait for ID
        ldi r17, SRX_ID
        sts serialmode, r17
        rjmp rx_done

        ;================================================================
        ;Check for target ID
rx_notwait: cpi r17, SRX_ID
        brne rx_notid

        ; If this isn't our target, wait for the next command frame
        cpi r16, MY_ID
        breq targeted
        rjmp rx_finish
targeted:
        ; If we HAVE been targeted, we next expect a command byte
        ldi r17, SRX_GETCMD
        sts serialmode, r17
        rjmp rx_done

        ;================================================================
        ;Get command byte
rx_notid:   cpi r17, SRX_GETCMD
        brne rx_notcmd

        ; COMMAND - Reset
```

```
          cpi r16, CHR_RESET
          brne cmd_notres
          lds r16, flags
          andi r16, ~(1 << FLAG_TESTMODE)
          sts flags, r16
          call clearscreen
          rjmp rx_finish

          ; COMMAND - Test mode
cmd_notres: cpi r16, CHR_TESTMODE
          brne cmd_nottest
          lds r16, flags
          ori r16, (1 << FLAG_TESTMODE)
          sts flags, r16
          rjmp rx_finish

          ; COMMAND - Blink on
cmd_nottest:   cpi r16, CHR_BLKON
          brne cmd_notblkon
          ldi r17, SRX_BLKON_COL
          sts serialmode, r17
          rjmp rx_done

          ; COMMAND - Blink off
cmd_notblkon:  cpi r16, CHR_BLKOFF
          brne cmd_notblkoff
          ldi r17, SRX_BLKOFF_COL
          sts serialmode, r17
          rjmp rx_done

          ; COMMAND - LED on
cmd_notblkoff: cpi r16, CHR_LEDON
          brne cmd_notledon
          ldi r17, SRX_ON_GETCOL
          sts serialmode, r17
          rjmp rx_done

          ; COMMAND - LED off
cmd_notledon:  cpi r16, CHR_LEDOFF
          brne cmd_notledon
          ldi r17, SRX_OFF_GETCOL
          sts serialmode, r17
          rjmp rx_done
```

```
cmd_notledoff:

        rjmp rx_finish


        ;================================================================
        ;Get column byte for BLINK ON
rx_notcmd:  cpi r17, SRX_BLKON_COL
        brne rx_notblkon_col
        subi r16, $41                    ; normalize column
        andi r16, $0F
        lsl r16                          ; multiply by 4
        lsl r16
        ldi r26, low(framedata)
        ldi r27, high(framedata)
        add r26, r16
        inc r26
        inc r26
        inc r26
        ld r16, X
        ori r16, $01
        st X, r16
        rjmp rx_finish


        ;================================================================
        ;Get column byte for BLINK OFF
rx_notblkon_col:cpi r17, SRX_BLKOFF_COL
        brne rx_notblkoff_col
        subi r16, $41                    ; normalize column
        andi r16, $0F
        lsl r16                          ; multiply by 4
        lsl r16
        ldi r26, low(framedata)
        ldi r27, high(framedata)
        add r26, r16
        inc r26
        inc r26
        inc r26
        ld r16, X
        andi r16, $fe
        st X, r16
        rjmp rx_finish
```

```
        ;================================================================
        ;Get column byte for LED ON
rx_notblkoff_col:cpi r17, SRX_ON_GETCOL
        brne rx_noton_col
        subi r16, $41              ; normalize column
        andi r16, $0F
        sts tmpcol, r16
        ldi r17, SRX_ON_GETROW
        sts serialmode, r17
        rjmp rx_done


        ;================================================================
        ;Get row byte for LED ON and switch LED on
rx_noton_col:  cpi r17, SRX_ON_GETROW
        brne rx_noton_row
        subi r16, $41              ; normalize row

        ; First calculate column RAM offset
        lds r17, tmpcol
        lsl r17                    ; multiply by 4
        lsl r17
        ldi r26, low(framedata)
        ldi r27, high(framedata)
        add r26, r17

        ; Does desired LED lie in the current byte?
        cpi r16, $08
        brsh on_b2                 ; no->

        ldi r17, $01
b1_1: cpi r16, $00
        breq on_done
        lsl r17
        dec r16
        rjmp b1_1

on_b2:inc r26                      ; Seek to next byte in framebuffer
        subi r16, $08              ; Chop off 8 from row#
        cpi r16, $08
        brsh on_b3

        ldi r17, $01
b2_1: cpi r16, $00
        breq on_done
        lsl r17
```

```
        dec r16
        rjmp b2_1

on_b3:inc r26                    ; Seek to next byte in framebuffer
        subi r16, $08            ; Chop off 8 from row#
        cpi r16, $08
        brsh on_b4

        ldi r17, $01
b3_1:  cpi r16, $00
        breq on_done
        lsl r17
        dec r16
        rjmp b3_1

        ; Rows Y and Z take special handling.
on_b4:inc r26
        subi r16, $08
        cpi r16, $00
        breq on_ry
        cpi r16, $01
        breq on_rz

        rjmp rx_finish           ; Column out of range.

on_ry:      ldi r17, $40
        rjmp on_done

on_rz:      ldi r17, $80
        rjmp on_done

        ; Finally we've finished this fandango and we write the video
byte
on_done: ld r16, X
        or r16, r17
        st X, r16
        rjmp rx_finish


        ;================================================================
        ;Get column byte for LED OFF
rx_noton_row:  cpi r17, SRX_OFF_GETCOL
        brne rx_notoff_col
        subi r16, $41            ; normalize column
        andi r16, $0F
        sts tmpcol, r16
```

```
            ldi  r17, SRX_OFF_GETROW
            sts  serialmode, r17
            rjmp rx_done


            ;===============================================================
            ;Get row byte for LED OFF and switch LED off
rx_notoff_col: cpi r17, SRX_OFF_GETROW
            brne rx_notoff_row
            subi r16, $41            ; normalize row

            ; First calculate column RAM offset
            lds  r17, tmpcol
            lsl  r17                 ; multiply by 4
            lsl  r17
            ldi  r26, low(framedata)
            ldi  r27, high(framedata)
            add  r26, r17

            ; Does desired LED lie in the current byte?
            cpi  r16, $08
            brsh off_b2              ; no->

            ldi  r17, $01
ob1_l:      cpi  r16, $00
            breq off_done
            lsl  r17
            dec  r16
            rjmp ob1_l

off_b2:     inc  r26                 ; Seek to next byte in framebuffer
            subi r16, $08            ; Chop off 8 from row#
            cpi  r16, $08
            brsh off_b3

            ldi  r17, $01
ob2_l:      cpi  r16, $00
            breq off_done
            lsl  r17
            dec  r16
            rjmp ob2_l

off_b3:     inc  r26                 ; Seek to next byte in framebuffer
            subi r16, $08            ; Chop off 8 from row#
            cpi  r16, $08
            brsh off_b4
```

```
        ldi r17, $01
ob3_l:      cpi r16, $00
        breq off_done
        lsl r17
        dec r16
        rjmp ob3_l

        ; Rows Y and Z take special handling.
off_b4:     inc r26
        subi r16, $08
        cpi r16, $00
        breq off_ry
        cpi r16, $01
        breq off_rz

        rjmp rx_finish          ; Column out of range.

off_ry:     ldi r17, $40
        rjmp off_done

off_rz:     ldi r17, $80
        rjmp off_done

        ; Finally we've finished this fandango and we write the video byte
off_done:   ld r16, X
        ldi r18, $FF
        eor r17, r18
        and r16, r17
        st X, r16
        rjmp rx_finish


rx_notoff_row:

        rjmp rx_done

        ; Finish transaction, return to SRX_WAIT
rx_finish:  ldi r17, SRX_WAIT
        sts serialmode, r17

        ; Clean up and exit ISR
rx_done:
        out SREG, r0
        pop r0
        pop r16
        pop r17
```

```
      pop r18
      pop r26
      pop r27
      reti
```

Video refresh functionality, including blinking of rows for which the blink bit is set, is handled in the timer 0 ISR:

```
;=======================================================================
; ISR
; Timer/counter 0 overflow
tc0_overflow:
      push r27
      push r26
      push r21
      push r20
      push r19
      push r18
      push r17
      push r16
      push r0
      in r0, SREG

      ; Turn off all column drivers
      ldi r16, $00
      out PORTA, r17
      nop
      ldi r16, $03    ; load zeros into U2, U3
      out PORTB, r16
      nop
      ldi r16, $00
      out PORTB, r16 ; disable load-enable for U2, U3

      ; Point X at framebuffer data
      lds r26, frameptr_lo
      lds r27, frameptr_hi

      ; Load data onto row drivers
      ld r18, X+
      out PORTA, r18
      nop
      ldi r19, $04
      out PORTB, r19
      nop
```

```
        ldi r19, $00
        out PORTB, r19
        nop

        ld r18, X+
        out PORTA, r18
        nop
        ldi r19, $08
        out PORTB, r19
        nop
        ldi r19, $00
        out PORTB, r19
        nop

        ld r18, X+
        out PORTC, r18

        ld r18, X+
        mov r21, r18
        andi r18, $C0
        out PORTB, r18

        mov r20, r18

        ; Check blink mode for this row
        andi r21, $01
        breq noblink

        lds r21, flags
        andi r21, (1 << FLAG_BLINK)
        brne noblink
        rjmp column_done

        ; Enable column driver
noblink: lds r18, currentline

        sbrc r18, 3    ; If the column modulo 16 is >= 8...
        rjmp sc_hi     ; ->

        ori r20, $01   ; Low bits are strobed with LE0
        rjmp sc_calc

sc_hi:ori r20, $02    ; High bits are strobed with LE1
```

```
sc_calc: andi r18, $07  ; mask off high bit of column#

        ; Shift left (R1) times to get the desired column latch byte
        ldi r17, $01
sc_c_lp: cpi r18, 0
        breq sc_c_done
        lsl r17
        dec r18
        rjmp sc_c_lp

sc_c_done:
        out PORTA, r17 ; put latch byte on PORTA
        nop
        nop
        out PORTB, r20
        nop
        nop
        andi r20, $C0
        out PORTB, r20

        ; Increment column
column_done:    lds r18, currentline
        inc r18
        andi r18, $0F
        sts currentline, r18

        ; Have we wrapped around to column 0?
        cpi r18, $00
        brne rfsh_cont

        ; Yes! Reset pointer and update blink count
        ldi r26, low(framedata)
        ldi r27, high(framedata)
        lds r18, framecounter
        inc r18

        ; See if we've had a blink overflow
        cpi r18, BLINK_RATE
        brne blink_ok

        lds r18, flags
        ldi r19, (1 << FLAG_BLINK)
        eor r18, r19
        sts flags, r18
```

```
        ldi r18, $00
blink_ok:
        sts framecounter, r18

rfsh_cont:

        ; store new frame pointer
        sts frameptr_lo, r26
        sts frameptr_hi, r27

        ; Force $80 into timer register to double frame-rate
        ldi r18, $80
        out TCNT0, r18

        out SREG, r0
        pop r0
        pop r16
        pop r17
        pop r18
        pop r19
        pop r20
        pop r21
        pop r26
        pop r27
        reti
```

This is the only project here that includes significant functionality in the main loop. This loop polls the test-mode bit, and when it is set by the serial ISR, the main loop calls the test-mode function. Test mode is provided mainly so the user can verify that all LEDs are wired and functioning correctly. In test mode, each entire column is illuminated sequentially (0–15), then each entire row is illuminated sequentially (A–Z). Any miswired or shorted row/column lines will become immediately apparent.

```
;=======================================================================
; Main program loop. Most of the functionality is actually in ISRs.
mainloop:
        lds r16, flags
        sbrs r16, FLAG_TESTMODE
        rjmp ml_nottest
        call testmode
        call clearscreen
```

```
ml_nottest:
      rjmp mainloop


;======================================================================
; Test mode
; Destroys R16, R19, R26, R27
testmode:
      ; PHASE 1
      ; Walk a line of FFs across the (16) columns
      call clearscreen
      ldi r18, $10

      ; Point X at start of framebuffer data
      ldi r26, low(framedata)
      ldi r27, high(framedata)
      ldi r20, $FF
      ldi r21, $C0

w_f_0:
      lds r16, flags
      sbrs r16, FLAG_TESTMODE
      ret

      lds r16, framecounter
      cpi r16, $00
      brne w_f_0

      ; store new data
      call clearscreen
      st x+, r20
      st x+, r20
      st x+, r20
      st x+, r21

w_f_1:
      lds r16, flags
      sbrs r16, FLAG_TESTMODE
      ret

      lds r16, framecounter
      cpi r16, BLINK_RATE / 2
      brne w_f_1

      dec r18
      brne w_f_0
```

```
                  ; PHASE 2
                  ; Walk a line of FFs down the (26) rows
                  call clearscreen
                  ldi r18, 26

                  ; Initialize data
                  ldi r28, $01
                  ldi r29, $00
                  ldi r30, $00
                  ldi r31, $00

w_r_0:
                  lds r16, flags
                  sbrs r16, FLAG_TESTMODE
                  ret

                  lds r16, framecounter
                  cpi r16, $00
                  brne w_r_0

                  ; store new data
                  ldi r26, low(framedata)
                  ldi r27, high(framedata)
                  ldi r25, $10
s_c_lp:     st x+, r28
                  st x+, r29
                  st x+, r30
                  st x+, r31
                  dec r25
                  brne s_c_lp

w_r_1:
                  lds r16, flags
                  sbrs r16, FLAG_TESTMODE
                  ret

                  lds r16, framecounter
                  cpi r16, BLINK_RATE / 2
                  brne w_r_1

                  ; shift "on-bit" one left through all 26 bits
                  cpi r28, $00
                  breq t_b1
                  lsl r28
                  brne t_done
```

```
        ldi r29, $01
        rjmp t_done

t_b1:   cpi r29, $00
        breq t_b2
        lsl r29
        brne t_done
        ldi r30, $01
        rjmp t_done

t_b2:   cpi r30, $00
        breq t_b3
        lsl r30
        brne t_done
        ldi r31, $40
        rjmp t_done

t_b3:   lsl r31

t_done:
        dec r18
        brne w_r_0

        rjmp testmode
```

At the very end of the code, we have a couple of miscellaneous subroutines to clear video RAM, and to reset all the output latches:

```
;=====================================================================
; SUBROUTINE
; Clear ALL framebuffer RAM
clearscreen:
        push r27
        push r26
        push r18
        push r16

        ldi r18, $00
        ldi r16, (4*16)
        ldi r26, low(framedata)
        ldi r27, high(framedata)

clslp:     st x+, r18
        dec r16
```

```
        brne clslp

        pop r16
        pop r18
        pop r26
        pop r27
        ret



;====================================================================
; SUBROUTINE
; Sets all output latches to LOW state (ie turn off row/col drivers)
clear_outputs:
        push r16

        ; Set output drivers for ports A,B,C low
        ldi r16, $00
        out PORTA, r16
        out PORTB, r16
        out PORTC, r16

        ; Latch zeros into U2,U3,U4,U5
        ldi r16, $0f
        out PORTB, r16
        nop
        ldi r16, $00
        out PORTB, r16

        pop r16
        ret
```

CHAPTER 4

The Linux-Based Controller (A Soft Task)

4.1 A Brief Introduction to Embedding Linux on PC Hardware

Before we start building an "embedded" distribution of Linux for our target platform, we need to formalize our goals for this system component. The requirements we define at this point will guide us in selecting our Linux components and determining how we construct the installation.

- Almost any embedded application needs to have turnkey characteristics; it needs to start a specific program at power-on and continue executing until power-down. Interactive startup events (for example, "press a key to continue" or "please login now") should be optional or nonexistent.

- The time required between power-on and full system functionality should be minimized.

- Unnecessary background tasks reduce overall performance, can only have a negative impact on reliability, and may even introduce other difficulties such as increased system power consumption or subtle security vulnerabilities. Therefore, any software modules and interface layers not essential to the actual application should be pruned out.

- Storage space and RAM are both usually going to be constrained. It is therefore desirable to configure installed software for the minimum possible "disk" and memory usage. (I put quotation marks around "disk" because the nonvolatile boot medium we use may in fact be some nonrotating storage device, such as flash memory).

- The major reason for bringing a large, complex operating system into our project is to facilitate the integration of off-the-shelf peripherals. Therefore, we need to select our various Linux software component versions for compatibility with the largest range of common consumer-grade hardware.

In this book, we will concentrate on kernel version 2.4.24, which is included on the CD-ROM for your convenience. This is not the most recent kernel version available at the time of writing; nor is it the version most commonly associated with embedded Linux applications. The reason I have selected it is because it is the most recent *stable* version in the 2.4.x version stream[18]. The previous major release (2.2.x) is very widely used in embedded applications today (particularly non-x86 Linux applications; ARM, MIPS, PowerPC, and so forth), but it is a less-than-ideal choice for us because the majority of ongoing maintenance work is currently aimed at 2.4.x and later kernels.

The latest 2.6.x kernels have greatly reduced interrupt latency (vs. 2.4.x) and other features highly desirable in embedded systems, but there have been some fairly major structural changes in these kernels. As a result, numerous third-party drivers (e.g., LIRC and the drivers for Atmel-based USB WLAN adapters, to take two random but personally important examples) cannot, at the time of writing, be built for 2.6. In order to make this text as widely applicable as possible, and to avoid getting bogged down in descriptions of how to patch various drivers to work with the 2.6 kernel tree, I have chosen to ignore these newer kernels. You should be aware that 2.6 is the way of the future; please refer to the companion web site for this book, *http://shoestring2.zws.com/*, where I will post notes and comments on embedding kernel 2.6 in the context of this text.

In the next few sections of this chapter, I will describe how to build a bootable filesystem on a CompactFlash card, ready to start your embedded application automatically at power-on. The exact same steps—perhaps with very minor modifications such as different target device name—can be used to prepare a variety of bootable media, including ZIP disks, USB "pen disks," and other miscellaneous removable me-

---

[18] This is constantly changing. As this book was being reviewed, kernel 2.4.26 was released. Although more than likely everything I've written here will "just work" with 2.4.26, I thought it inadvisable to take the risk of rewriting this text at such a late date.

dia (as long as your target system has BIOS support for booting from these devices). I'll also show you how to build a bootable system restore CD-ROM that can be used to dump an entire system load onto a blank hard disk and make it bootable. If you're shipping turnkey Linux systems that must boot off a hard drive, a system restore CD is virtually mandatory—it is a robust way of ensuring that even in a worst-case scenario, you won't have to go out for field service calls or waste money shipping corrupted units to and fro.

It is important to stress that the methods and aims we will describe in this chapter are not precisely the same as those for building traditional monolithic single-processor systems running Linux. To take one example in particular, the run-time library we will be using to interface application code to the Linux kernel is the "full" version of glibc. Embedded applications would not normally use this very heavy run-time; they would normally use a cut-down runtime such as uclibc. Deeply embedded projects would also be considerably more rigorous about eliminating cruft than the text you're about to read. The emphasis I'm pushing is ease of assembly and, even more importantly, good compatibility with "desktop" Linux so you don't have to delve too deeply into the mysteries of guru-level embedded Linux magic. If you want more detail about the subject matter of this chapter, an excellent place to start your further reading is the Linux from Scratch project at *http://www.linuxfromscratch.org/*. The companion book to that web site is "Linux from Scratch, Version 4.1," ISBN 0-9659575-6-X, published by Clearly Open. (That ISBN is for the book with companion CD-ROM—a version without the CD-ROM is also available). Linux from Scratch covers a lot more territory than I will here, since they are concerned with providing you with enough information to build distributions with the complexity level of a complete desktop system.

4.2 Configuring the Development System and Creating Our Custom Kernel

One very useful side-effect of using a PC as the central system controller is that you can run all the development tools directly on the target hardware (or possibly a slightly expanded version of that hardware), eliminating the need for a complex cross-development system. The main reason I advocate doing this is because bitter experience has taught me that the build and install process for some third-party

Linux drivers has only been tested in the "vanilla" case, that is, building and installing into the currently active system. Automatic configuration scripts that test for the avail-ability of specific libraries and/or hardware features (for example, CPU instruction set extensions such as MMX, SSE, 3DNow! and so on) will also require coercion if you're running them on a system other than the target hardware. Debugging these sorts of issues with cross-compiling and manual installation of these components is a waste of your time. Thus, I suggest that you start by building your SBC into a fully-configured PC by adding a keyboard, mouse, hard disk and CD-ROM drive,[19] and perform all your kernel and utility compilation directly on the SBC. You should install Fedora Core with a custom configuration; don't install XFree86[20] or associated frippery (graphical In-ternet tools, games, GNOME, KDE, etc.) unless you specifically want them. Otherwise, just install the core operating system, development tools, and kernel development. If you want an exact list of what to select in the install dialogs, here is a painstakingly exact description of how to navigate the installation process (note: this information is correct for Fedora Core 1 "Yarrow." The reason I am spelling it out in such painstaking detail is so that you can be certain you're working with the exact same system configu-ration I was working with when I wrote and tested the code in this book):

- Boot off the CD and type `linux text` to use the text-mode installer. Vagaries of the Geode graphics subsystem mean that the graphical installer probably will not work correctly.

- Choose "Skip" to skip testing of installation media, and select "OK" in the welcome dialog.

- Select "English," then "us". Choose the type of mouse, if any, connected to your system. If you selected a serial mouse, indicate the port to which it is at-tached (probably /dev/ttyS0).

---

[19] You can actually buy some of the SBCs on our recommended hardware list preassembled in a box with a power supply, CD-ROM drive and hard disk. However, it's *much* cheaper—as much as 50% cheaper – to put it together yourself. The system fits very elegantly into a housing from an external 5.25″ disk or tape drive, if you happen to have one lying about.

[20] If you install XFree86, the system will probably not boot correctly unless you first go into CMOS setup and ensure that the LCD resolution is set for 1024 × 768 and video memory size is set for 4.0 MB. With certain older BIOS versions, you need to do these steps even if you're not using the digital-output LCD port on the board.

- Accept the default monitor settings, and select "Proceed" in the warning dialog that follows.

- If you're installing on a hard disk that already contains a Linux installation, Fedora will ask you if you want to upgrade the existing installation, or reinstall. Select "Reinstall".

- Select the "Custom" installation type.

- Select "Disk Druid" partitioning. The next dialog you see will be Disk Druid. How you partition your system is up to your own preference, but for this sort of application I normally create a 256 MB swap partition and use the remainder of the disk as the root partition. (No, this is not normally regarded as best practice; but unless you have special requirements, it's the simplest partitioning scheme for this single-function development system). Always force your partitions to be primary partitions. After partitioning, you'll probably get a dialog warning that swap space is going to be turned on immediately; just select "OK".

- Select "Use GRUB Boot Loader". In the next four dialogs (all of which are titled Boot Loader Configuration), you should not need to specify any special parameters; just select OK. This will install GRUB in the MBR, with no special options or password protection.

- You now need to configure the built-in RTL8139 Ethernet adapter. Since the configuration we're setting up now is only used for development, I suggest you leave it at the default settings, which activate the interface automatically on boot and attempt to acquire an IP address using DHCP.

- In the next dialog (Hostname Configuration), you can either leave the settings at defaults, or manually enter a hostname for this system.

- Since this is a development system that should already be isolated from network attacks, select "No firewall" in the Firewall dialog. Select "Proceed" in the warning dialog that will follow.

- In the next dialog (Language Support), just select "OK".

■ In the Time Zone Selection dialog, select the timezone you're working in and select "OK".

■ Choose a root password and enter it twice in the Root Password dialog.

■ There will now be a *LONG* pause while the installer analyzes the list of installable packages. The system isn't crashed; just be patient. Once the Package Group Selection dialog comes up, select only the following packages (deselect all others): "Editors," "Text-based Internet," "Windows File Server," "Network Servers," "Development Tools," "Kernel Development," and "System Tools." Note that in some cases, when you press Space to enable an option, the system will become unresponsive for up to 30 seconds; this is the install script analyzing the packages you've selected. After you select OK in the Package Group Selection dialog, select OK in the "Installation to begin" dialog, select "Continue" in the Required Install Media dialog, and the install process will commence. At the end of the installation, select "Reboot" and you're done.

The process I just described yields a system on which you can build and test your embedded kernel directly. You can also assemble your bootable CompactFlash image and write it to a CF card on this same piece of hardware, then test booting off it simply by altering the CMOS settings on the SBC—which is a big time-saver while you're debugging CompactFlash startup issues. However, if you're unwilling or unable to use the target SBC as your development system (for instance, if it's physically installed into a piece of equipment and it's difficult to attach development peripherals), there is an alternative method you can use. This method can be summarized as follows:

■ Take the source for the kernel version you intend to use on the SBC, and configure and build it for your **development system.**

■ Install the kernel and modules on your development system, and modify your bootloader so that it loads this kernel. For the sake of this discussion, I'll assume that you named the active, booting kernel image "/boot/bzimage-2.4.24". Edit your bootloader's configuration appropriately.

■ Reboot so that you're running the same kernel version that you intend to use on the SBC.

- Archive the kernel and its configuration. You can back up the important parts of the active Linux environment simply with the command:

```
tar cvfz /activekernel.tar.gz /lib/modules/* /boot/bzImage-2.4.24
/etc/modules.conf
```

- Clean up the kernel and module directories by executing the commands `make clean` in the kernel source directory and `rm -rf /lib/modules/2.4.24` to remove the active kernel modules.

- Configure the kernel for your SBC, build and install it. If you're using the Advantech board or a compatible device, you can use the kernel configuration I have supplied on the CD-ROM. **Don't use the same path for the kernel image file (bzImage) as you're using for the real kernel that boots your development system, or you might not be able to start the development system easily if something goes wrong and you have to reboot.** I suggest you install the SBC-specific kernel as /boot/bzImage-sbc. If you're using the materials supplied on the CD-ROM, here's the exact set of commands you'll need to execute:

```
cd /usr/src
tar zxvf /mnt/cdrom/linux/linux-2.4.24.tar.gz
cd linux-2.4.24
cp /mnt/cdrom/linux/geode-config .config
make dep ; make bzImage ; make modules ; make modules_install
cp arch/i386/boot/bzImage /boot/bzImage-sbc
```

- Configure, build and install any additional drivers you need (for example, LIRC, mentioned later in this text, drivers for WLAN devices, and so on).

- Archive the SBC's kernel configuration:

```
tar cvfz /sbckernel.tar.gz /lib/modules/* /boot/bzImage-sbc /
etc/modules.conf
```

- **Immediately** after you finish archiving up the completed kernel installation for the SBC, restore your development system's bootable kernel set with the following commands:

```
rm -rf /lib/modules/2.4.24
cd /
tar zxvf activekernel.tar.gz
```

This way of doing things is not quite as desirable as development on the real hardware, because you still have to hand-tweak configuration files for applications and libraries that auto-configure for installed hardware. You may also have some (usually very minor) problems with module dependencies, requiring manual editing of the modules.dep file, and you obviously can't actually test the kernel and drivers on your alien development hardware. However, this system does allow you to build the kernel and all the installable kernel modules in a reasonably robust and simple way. Note, by the way, that many installable driver packages place daemons, libraries, utilities and other files in various directories other than /lib/modules/kernelversion, and some of these packages modify startup scripts to load daemons automatically on boot. You'll have to identify and copy these extra files and script modifications across to the SBC by hand.

The reason I advocate this somewhat tortuous process in lieu of performing a true "cross-build" (coercing install scripts to use a kernel build tree different from the active one reported by uname(1)) is because I have frequently encountered bugs in the scripts and makefiles for various third-party kernel drivers and other software, which only manifest if you are installing to a destination other than the current (booted) kernel directory. It would appear that most such drivers are developed and tested under the assumption that you will be building and installing the driver into the currently active environment. Rather than analyzing and testing the configure scripts and makefiles for each third-party driver on a case-by-case basis (and perhaps risking that a critical component won't be transferred correctly to your target system), it is much simpler simply to build "live." If you can't or won't build on a truly live installation on the real target hardware, the next best thing is to simulate, on your development PC, the same kernel environment that will be running on the SBC.

All of this complexity is avoided if you build your Linux environment directly on the target platform, so I reiterate: if at all possible, build up your SBC into a semi-complete, usable system so you can build your software directly on the real hardware. It will save you a vast amount of time and frustration.

4.3 The Linux Boot Process—Creating a Bootable CompactFlash Card

Let's begin this section where your embedded application begins: with the power-on boot process of the SBC and Linux. At power-on or hard reset, the BIOS probes attached hardware and builds a list of bootable devices. It refers to a boot priority table in ROM or nonvolatile CMOS RAM and scans available boot devices in the order specified in that priority table. For each device, the first sector on the disk is read into RAM and scanned for a boot signature. If the signature is found, the BIOS assumes the disk is bootable, and jumps into the bootsector code copy in RAM. This code, installed by the operating system's partitioning or formatting utility, loads the remainder of the operating system, possibly in several stages. In a Linux system, this tiny snippet of bootstrap code is either a header prepended to the kernel itself (such a configuration has just enough intelligence to load the remainder of the kernel from consecutive sectors and jump into it), or the first stage of a bootloader program that can gather configuration information from various sources and handle a more complicated load process. In the case of a Linux installation, the bootloader loads at least the kernel image, and perhaps also an initial RAMdisk (initrd) image into memory before passing control to the kernel. The initial RAMdisk image is intended to contain modules, scripts and other data that may be required by the kernel before the real root filesystem can be mounted. It is normal for the on-disk copy of the image to be compressed with gzip(1). Once the initrd image is loaded into RAM, decompressed and mounted, the kernel spawns /sbin/init, which continues the boot process[21].

There are a myriad different possible ways to organize the boot process of a Linux system. When booting from a hard disk, it is usual to employ either LILO (LInux LOader) or grub—modern Linux distributions tend to favor grub. Both of these programs are powerful boot-manager applications capable of loading different, user-selected operating systems at boot time, offering simple password protection, operating system selection menus on the local console or a serial port, and numerous

---

[21] Please note that this is a *considerable* simplification. There is actually a handover point where the initial RAMdisk is destroyed and the real root filesystem is mounted, and there are several other possible steps and forks in the boot process. We won't be dealing with all these options in our application, because we will be running permanently out of the initial RAMdisk.

other features. They're not routinely necessary for an embedded application, though it is possible to use them on media other than hard disks if you want to. One application where this sort of boot menu can be handy is if you want to provide some kind of emergency recovery facility where the user can boot the system in emergency mode and perform some sort of recovery or diagnostic operations. However, what we are trying to do (make a bootable CompactFlash card or CD-ROM) is more akin to preparing a bootable Linux floppy disk than a full-blown hard disk boot. In fact, in the case of the bootable CD-ROM, it's almost exactly like a floppy disk.

The ancient and still somewhat documented method of preparing a bootable Linux kernel floppy is to patch (some might say, "mercilessly hack") the kernel image using the ancient and deprecated rdev(8) utility, then use the dd(1) utility to write the kernel image itself (bzImage) directly to a floppy disk, possibly followed by an initial RAMdisk image. If you're searching the Internet for information on how to build bootable floppy disks, this method is likely to crop up high in your search results, but it's a red herring. There are a few problems with it, and the main problem is this: After the kernel code starts executing, BIOS functionality is more or less completely preempted. This means that if you want the kernel to be able to load an initial RAMdisk, it either has to be on an actual, separate floppy disk, or it has to be small enough to fit right after the kernel image on the first floppy disk. As kernel sizes have increased, it has become close to impossible to fit everything needed in a kernel (typically as much as 700K) and initial root filesystem onto a single floppy disk. Thus, it's unreasonably difficult to make, say, a bootable CD-ROM using this method, because as soon as the kernel starts executing and goes looking for its initial RAMdisk image (on a second floppy disk), it will try to load it off actual floppy sectors instead of going through the BIOS's emulated floppy-on-CD-ROM layer, and the boot process will fail.

Another way of constructing a bootable medium is to format your device, or a portion of it, as FAT, make it bootable with your choice of DOS versions, and create an AUTOEXEC.BAT that uses the LOADLIN utility to load a kernel and initial RAMdisk. LOADLIN was popular for making "Start Linux now!" icons on systems running Windows 3.x through 98, because it allows you to start Linux directly from a DOS-based operating system without needing to reboot. It also allows you to do tricky things like having Linux on a secondary drive (perhaps a removable drive) with no footprint at all on your main Windows drive—and furthermore, you don't need to worry about Windows or a bootsector virus accidentally wiping out your LILO or grub installation. As long as you can boot Windows, you can boot Linux. The downside to the DOS + LOADLIN method of doing things, from our perspective, is that it requires a whole throwaway operating system kernel just to act as a bootloader. This is an unnecessary waste of flash space, and it also violates the universal common sense design rule of eliminating software layers not actually necessary for the application at hand.

As a result, we're going to use the SYSLINUX bootloader in our system. SYSLINUX is a beautifully small and simple piece of software. It consists of an installer utility (syslinux under Linux, SYSLINUX.EXE under DOS/Windows), a small loader module, and a kernel and initial RAMdisk image that you supply. The function of the syslinux utility is basically the same as the "/s" switch to the DOS FORMAT command; it makes the target media bootable. Instead of installing the MS-DOS IO.SYS, MSDOS.SYS and COMMAND.COM files, however, syslinux copies across its own loader module, which is called LDLINUX.SYS. The loader module examines and executes the configuration file SYSLINUX.CFG, which is a plain text file (in the root directory of the boot medium) instructing SYSLINUX which kernel to load, what options to pass it, and what initial RAMdisk to load.

Once the kernel and initial RAMdisk are loaded, a "normal" Linux distribution will mount the real root filesystem (usually a hard disk—sometimes an NFS sharepoint or some other sort of device) and discard the contents of the initial RAMdisk. In our embedded application, however, we're going to run with a RAM-based root filesystem. This minimizes wear on the flash media, and also improves system reliability (since we're never writing any changes to the on-disk copy of our root filesystem, it is always a known-good start point). Whether running the root filesystem out of RAM or from some other device, the next thing the kernel does is to run /sbin/init to process system startup scripts and bring up whatever network connections, device drivers and daemons you wish to have running on the system.

Figure 4-1 summarizes just a few of the possible paths we could travese to start up our Linux system, including the methods described above. We are going to follow the path that goes from BIOS to SYSLINUX to kernel, and then directly to /sbin/init.

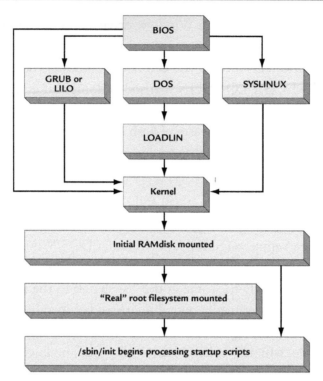

Figure 4-1: Some possible Linux boot paths

By the way, this is an appropriate moment to point out why it's extremely desirable to use FAT format on your bootable CompactFlash card. The most obvious reason is that Windows can read and write it directly. This makes field-upgrades of firmware much easier than they would be if you were using a Linux-specific filesystem. Instead of having to write, or at least deploy, a disk-imaging utility to your users, you can simply email them new kernel and initrd files, and tell them to use a regular USB CompactFlash reader to copy the new files onto their existing card. Another related benefit is that any logged data your embedded appliance cares to write to the card will be accessible directly by Windows users, without requiring any special drivers.

A more subtle reason for using FAT is that, by and large, removable flash media are designed with the assumption that FAT will be the filesystem in use. The exact implications of this are rather "gray" (i.e., implementation- and situation-specific), but for example, many NAND flash devices—primarily SmartMedia, but also some CompactFlash cards—have a low-level format specification that specifies a 1:1 correspondence between FAT clusters and erasable flash blocks. Using the correct FAT format for your particular card size results in far fewer erase-write cycles, thereby extending the operational life of the card.

Note: If you are using extremely small CompactFlash media (less than 16 MB in size), you may encounter problems using mkfs (or, more strictly speaking, mkfs. msdos) to format the card; you will receive an error that the volume is too small to create a filesystem. I have not found a fully satisfactory workaround for this problem. The easiest way to avoid the problem is to use some other appliance or operating system—Windows or a digital camera, for instance—to format the card. You can also use the mformat(1) command, from the mtools package, to format the card properly.

Now that I've outlined to you the combination of software we intend to use, let's actually install it on the card. I'll assume you are doing this on an actual PCM-5820 or similar SBC, where the CompactFlash slot is wired in true-IDE mode, configured as the master device and connected to the secondary IDE interface; i.e., at /dev/hdc[22]. I'll also assume that the card is already partitioned and formatted as FAT12, FAT16 or FAT32 (this will be true of a brand-new card out of the box). The sequence of steps to follow is:

1. Install SYSLINUX on the card: `syslinux /dev/hdc1`.

2. Mount the card so we can access it: `mount /dev/hdc1 /mnt`.

3. Copy the Linux kernel across: `cp /boot/bzImage-sbc /mnt/LINUX`.

4. Create a text configuration file called SYSLINUX.CFG in the root of the card (/mnt/SYSLINUX.CFG). A suggested configuration file might be simply:

   `DEFAULT LINUX initrd=INITRD.IMG root=/dev/ram`

   You can also add other kernel parameters to this file—for instance, you could redirect the console to a serial port (console=/dev/ttyS0), specify a different video startup mode (vga=*mode-number*), and so on.

5. Unmount the card: umount /mnt.

6. The card is now bootable (though not usefully so—read on!).

---

[22] Keep in mind that true-IDE mode does not support hot-swapping. You must power off the system before inserting or removing CompactFlash cards.

By the way, the filename capitalization I've used isn't mandatory—SYSLINUX won't care—but I've explicitly used capitals here in case you happen to mount the card with VFAT long filename support, which also includes support for mixed-case filenames via some slightly tortured semantics.

We're half-way to a hand-made standalone Linux system! Proceed to the next section for the other half of the puzzle. If you want to test what you have created so far, you can disable the hard drive in your SBC's CMOS settings and let the system boot off the CompactFlash slot. You'll see the usual kernel startup messages, followed by an error stating that the root filesystem couldn't be mounted.

4.4 Creating a Root Filesystem for our Embedded System

At the end of the previous section, we had the system successfully booting off the CompactFlash card, at least as far as loading the kernel. This isn't enough, though—in order to do anything useful, we also need to build at least a basic root filesystem to live in the RAM disk.

In order to generate the root filesystem image, we need to be running with a kernel that supports the loopback device, so that we can mount and manipulate disk image files just as if they were physical storage devices. The kernel configuration I supplied for you on the CD-ROM has this support enabled. If you're using a different computer to build your Linux software, you may need to recompile your kernel and add loopback support; you'll find it under "Block devices" in 2.4.x kernels, or in "Device Drivers—Block devices" in 2.6.x.

Let's first create an empty file called /initrd.img, exactly 8 MB in size, to contain our root filesystem image. We use dd(1) to achieve this (dd is one of several *NIX commands that are apparently officially depreciated, but without which people don't seem to be able to live):

```
dd if=/dev/zero of=/initrd.img bs=1k count=8192
```

Bear in mind that the entire disk image will be in RAM—so the larger you make it, the less heap will be available to your applications. Also consider that the version of the image file *stored on the boot medium* will be compressed with gzip. The kernel will load the entire image into RAM and decompress it before mounting it. The larger you make your image file, the slower the boot process—though initrd decompression isn't usually a major time hog during boot. In any case, I chose the size 8 MB because it's a convenient number, and because I happen to know that all the items I want to include will fit in there with plenty of space left over. If you need to fine-tune this more, some trial and error will be necessary.

With these considerations in mind, we next turn this empty file-o-nulls into a valid ext2fs filesystem image using mke2fs(8):

```
echo y | mke2fs -m 0 -N 2000 /initrd.img
```

Note the options here. By default, 5% of every ext2 volume is reserved for the superuser. We have no need of such multi-user frippery, so we turn that option off with -m 0. This isn't absolutely necessary, but it's a neatness issue, like securing interconnect cables inside your device with zip-ties instead of leaving them flapping in the air. Secondly, and much more importantly, the ext2 filesystem is built on an inode table, the size of which is fixed at format-time. By default, mke2fs calculates the default number of inodes based on the volume size. For the tiny volumes you'll be working with, the default numbers are almost certainly not going to be big enough, which will mean you'll be unable to copy all the files you're going to need—not because there is insufficient disk space, but because there are insufficient inodes to describe the files. The –N 2000 parameter forces the creation of 2000 inodes, which should be enough for us. If you encounter strange errors when copying files into the image, reformat it with a larger –N parameter and try again.

To digress slightly, this issue can easily be put in a perspective appropriate to FAT mavens. It is very closely analogous to DOS's limitation that the root directory of a volume can only contain a fixed maximum number of entries, the number of which is set at format time (and is based on the cluster size—hence, indirectly, on the volume size). Attempt to create one too many files or subdirectories in the root directory of a FAT volume and you'll get an error message, the exact unintelligibility of which depends on the operating system you're running (Windows Explorer unhelpfully and inaccurately reports that the disk is full). This FAT misfeature can and does create technical support issues in embedded environments. It's also why appliances like digital cameras always store their files inside a subdirectory—because subdirectories can be expanded indefinitely, up to available disk space.

We've pulled together almost all the components we need. However, we're still missing a bunch of little utilities that are essential to creating a Linux distribution; /sbin/init, a shell, and so on. Although we could simply pull the appropriate utilities out of our active desktop Linux distribution, this isn't a very efficient or embedded-friendly way of doing things. A much better answer is to use the busybox tool, which is designed specifically to fill in these gaps for embedded environments. Busybox is found in a large majority of embedded Linux distributions; it is an all-in-one program that includes the functionality of several dozen standard Linux command-line utilities. Ingeniously, all these utilities are rolled into a single executable; you access the various goodies buried in the busybox program by creating multiple symlinks with different names. Busybox looks at _argv[0] and determines how to behave based on which symlink was used to invoke it. I have included busybox version 0.60.5 on the CD-ROM. Note that you can select which functional blocks are compiled into the main executable by editing the Config.h file in the busybox source directory. The configuration I used for this book is supplied for you on the CD-ROM as /linux/busy-box-config.h.

Let's start building the collection of files that will eventually comprise our root filesystem. (In the following discussion, I assume you will use /tmproot as the directory to assemble the new root filesystem). Before you start groaning at the amount of typing to be done, rest assured that I've done all the hard work for you—you can find the complete root filesystem in the file card-root.tar.gz in the root directory of this book's CD-ROM. Untar this archive and you will be left with a directory called "tmproot" containing all the files required for the root filesystem described next.

To create such a filesystem from scratch, the first thing we need to do is create a collection of symlinks to busybox within the root hierarchy, and copy the busybox executable itself. This is the kind of tedious job you only ever need to do once. Create a bin directory in /tmproot, and copy the busybox executable there. Here is a list of the items that need to be symlinked to it. Note that these paths are relative to the /tmproot directory—I suggest that you cd /tmproot/bin (or whatever directory you're assembling) then ln -s [filename] ./busybox, or for the items in /sbin, ln -s [filename] ../bin/busybox.

/bin/ash	/bin/head	/bin/swapoff
/bin/basename	/bin/id	/bin/swapon
/bin/cat	/bin/kill	/bin/sync
/bin/chgrp	/bin/killall	/bin/tail
/bin/chmod	/bin/linuxrc	/bin/tar
/bin/chown	/bin/ln	/bin/test
/bin/chroot	/bin/logger	/bin/touch
/bin/chvt	/bin/ls	/bin/true
/bin/clear	/bin/lsmod	/bin/tty
/bin/cp	/bin/mkdir	/bin/umount
/bin/cut	/bin/mknod	/bin/uname
/bin/date	/bin/mkswap	/bin/uniq
/bin/dd	/bin/more	/bin/uptime
/bin/df	/bin/mount	/bin/wc
/bin/dirname	/bin/mv	/bin/which
/bin/dmesg	/bin/pidof	/bin/whoami
/bin/du	/bin/poweroff	/bin/xargs
/bin/echo	/bin/ps	/bin/yes
/bin/env	/bin/pwd	/bin/zcat
/bin/false	/bin/reset	/sbin/init
/bin/find	/bin/rm	/sbin/klogd
/bin/free	/bin/rmdir	/sbin/modprobe
/bin/grep	/bin/sed	/sbin/reboot
/bin/gunzip	/bin/sh	/sbin/swapoff
/bin/gzip	/bin/sleep	/sbin/swapon
/bin/halt	/bin/sort	/sbin/syslogd

To support some applications we're going to talk about later, you should also copy across a few files from your active Linux installation: /bin/eject, /sbin/fdisk, /sbin/mke2fs, and /sbin/mkswap. More about those a little later.

The next thing we need to do is create a /dev directory containing entries for all the devices we're likely to use. (This step could be avoided by compiling devfs support into the kernel, but devfs is not terribly well-supported yet and in my humble opinion is best avoided until it is universally embraced). There are two ways to create your /dev directory: the tidy way, and the easy way. The tidy way is to make a list of all the devices you know you're going to use, and use mknod(1) to create exactly those entries—no more, no less. On a moderately complex system, however, this can be a REALLY tedious task, with a lot of trial-and-error testing later. The easy way is to cheat: cp -a the /dev directory from your development system into the filesystem image (`cp -a /dev /tmproot`), then delete the device entries you don't need from /tmproot/dev. It isn't necessary to be particularly rigorous in the latter operation—if you leave in something unwanted, at worst you'll be wasting a little disk and RAM space. Since our kernel is very general-purpose—I don't know what you want to do with it, so I included a lot of optional drivers—the easy way is better for us.

Although not absolutely necessary for a basic Linux system, we should also create a /proc mountpoint—many utilities try to get system status information out of /proc pseudofiles and won't work properly if the /proc pseudofilesystem isn't mounted.

The penultimate piece of magic we need to add is the necessary dynamic-link libraries to support busybox (and any other programs we are including on the boot medium). To discover what libraries are required by busybox, we first compile it, then use the ldd(1) utility to show its dependencies. This will tell us the following (the exact output you will get here depends on the Linux environment you used to build the program):

```
# ldd ./busybox
    libc.so.6 => /lib/tls/libc.so.6
    /lib/ld-linux.so.2 => /lib/ld-linux.so.2
```

So, we need to supply the libc.so.6 runtime library and the ld-linux.so.2 dynamic loader. Normally, libc.so.6 is a symlink to the "real" runtime library, and ld-linux.so.2 will be a symlink to the real loader. In Fedora Core 1, these will be libc-2.3.2.so and ld-2.3.2.so, respectively (both are located in /lib)[23]. So we can set up the environment for busybox using:

```
cp /lib/libc-2.3.2.so /tmproot
cp /lib/ld-2.3.2.so /tmproot
cd /tmproot
ln -s libc-2.3.2.so libc.so.6
ln -s ld-2.3.2.so ld-linux.so.2
```

You can use basically the same process to determine the library dependencies of any other programs you want to include in your little distribution. Of the programs I listed on p. 131, only mke2fs requires additional libraries over and above ld-linux.so and libc.so.

Note, however, that ldd is usually not the complete answer to transferring a program from your full desktop distribution to a limited embedded environment. For example, it won't tell you what temporary directories a program is trying to create, or where it's trying to write its log file, or where it tries to read configuration information. In a few rare and extraordinarily evil cases, you will also encounter programs that load dynamic-link libraries explicitly at runtime (one such program is passwd(1)). There's no automatic way to find out if the program does tricks of that sort.

To ascertain the details of more advanced cases like this, you should first refer to the documentation for the program you're trying to transfer. If you still can't get it working, the brute-force approach is to run the troublesome program inside strace(1) and log the result, for example with the commandline `strace myprogram 2>&1 |tee myprogram.out`. This will give you a very verbose logfile, which you should peruse looking for file open calls—you can infer from this information what directories and/or files the program is trying to access. The strace attack is really an approach of last resort, though—it's tedious, requires lots of analysis, and is not entirely error-proof.

---

[23] This technique allows you to upgrade the underlying library without breaking installed applications.

At this point, you have enough material copied across to test your copy of the soon-to-be root filesystem. Simply execute `chroot /tmproot /bin/ash -i` and, if everything has gone to plan, you'll see the Busybox shell prompt. The environment you're now in is pretty close to what your embedded system will see; anyway, it's close enough for you to be able to test shell scripts and some executables.

What you've assembled so far is enough for the system to be able to do useful work; the only thing we're missing is startup instructions to tell it what work we want it to do! These instructions are contained in three files: /etc/fstab, /etc/inittab and /etc/sysinit. I won't dwell on the format of fstab, which specifies the mountpoints for the filesystems in your environment; it's a minor piece of study and you can learn all you need to know from `man 5 fstab`. Let's start with inittab. This is the configuration file for init(8). The sample file I've provided is as follows:

```
::sysinit:/etc/sysinit
::respawn:/bin/ash
::ctrlaltdel:/sbin/reboot
::shutdown:/bin/umount -a -r
::shutdown:/sbin/swapoff -a
```

This tells init to run /etc/sysinit at system startup, to run /bin/ash after system startup and to respawn it automatically if it exits, and it also provides some clean-up functionality to handle the user pressing Ctrl-Alt-Del or shutting down the system some other way. Note that this file is in the correct format for the init variant built into busybox; due to minor syntactic differences, you generally can't take an inittab from a "normal" desktop distribution and use it with busybox. Also note that if you wanted this card to run your own application automatically at startup, you could specify the path to your executable instead of /bin/ash in the respawn line. For more information on the specifics of writing an inittab for the busybox implementation of init, refer to the busybox documentation.

Our /etc/sysinit contains the following:

```
#!/bin/ash
HOSTNAME=geode
mount -n -t proc /proc /proc
mount -n -o remount,rw /
#swapon -a -e
```

```
    >/etc/mtab
    mount -f /
    mount -f /proc
if ( cat /proc/cpuinfo | grep -s Geode >/dev/null ) ; then
    echo "Welcome to the Geode-based SBC."
else
    echo "Warning: This is not recognized as a Geode."
    echo "Some kernel features may not work correctly."
fi
```

This script is just an example of some of the housekeeping sort of functions you might need to carry out at system boot time; we set a friendly hostname for this machine, mount the /proc filesystem, then remount the root filesystem as read-write. Note the use of the –n parameter to mount—because at this point, the root filesystem is still readonly, and unless we specify –n, mount will try to update /etc/mtab and fail. After remounting the root filesystem, we might want to turn on swap, if our system has a swapfile (the example I've prepared assumes you don't. It's not a good idea to put a swapfile on a CompactFlash card or other flash device because you rapidly eat up the write endurance of the flash chips). After that, we clear out any ancient data that might be in /etc/mtab, then fake (with the –f switch) remounting / and /proc, so that mtab contains up-to-date information. Finally, I perform a little sanity test to show a welcome message appropriate for the machine.

With all these components prepared, we're now ready to copy the root filesystem into the image file we prepared for it. **Note a subtle pitfall lurking for the unwary:** The image file created with dd starts off full of highly-compressible all-zero data (this is why we use /dev/zero as the dummy data source to fill that image file during creation). If you manipulate the contents of a mounted image file using standard Linux file management functions, the image will gradually fill up with less-compressible file data droppings belonging to moved or deleted files. Eventually, you might be unable to fit the compressed version on your boot medium. (Remember how defragmenting a disk compressed with Drivespace/Doublespace used to increase the available space magically? Same principle). For this reason, when you are preparing a set of files to create a new initrd image file, even if you're just changing a few bytes from the previous version of the image, it's better if you create a new, all-zero image file, format it, and copy the complete set of up-to-date files into it. What I usually do is create a working directory on my hard drive, containing the complete root image, and write a simple shell script that creates a virgin image, formats it, copies everything into it, and compresses it ready for use. Such a script might look like this:

```
#!/bin/sh
cd /tmproot
dd if=/dev/zero of=/initrd.img bs=1k count=8192
echo y | mke2fs -m 0 -N 2000 /initrd.img
mount -o loop /initrd.img /mnt
cp -R * /mnt
umount /mnt
gzip /initrd.img24
```

Run the above script, or type the commands manually, and copy the /initrd.img.gz file thus generated into the root directory of your CompactFlash card with the name INITRD.IMG. Reboot your SBC and go into CMOS setup. Disable the hard disk or simply choose the CompactFlash slot as the primary boot device[25], and restart the system again. If everything went to plan, you should be greeted with a usable busybox shell prompt within less than 30 seconds.

4.5 Creating a Bootable Linux System-Restore CD-ROM Disc

Practically all modern IBM-PC compatible BIOSes (including the Award BIOS used on our SBC) support the ElTorito standard for bootable CD-ROMs. This standard places a special preamble area on the disk, which contains a bootable image file. The boot image is normally a sector-by-sector copy of a floppy disk, although it is legal to have a hard disk image there instead. After POST, if the system is configured to attempt booting off CD, the BIOS scans the disk in the attached CD-ROM drive for an ElTorito preamble. If this preamble is found, the BIOS redirects read and (doomed!) write requests for INT 13h unit 0 (drive A:) to the boot image area on the CD-ROM drive, and continues the boot process from the resultant virtual disk.

---

[24] Remember, gzip will automatically add a .gz extension to the compressed file.

[25] Unfortunately, on a few systems, you may need to physically disconnect the hard drive to force the system to probe further for a bootable device. This behavior is BIOS-specific. On the PCM-5820, simply setting the primary master hard disk type to None and the secondary master type to Auto will allow you to boot off the CompactFlash slot.

As far as your startup software is concerned (at least, as long as it uses BIOS services to read/write the disk), you just booted the system off a floppy.

In this section, I'll demonstrate one possible method of constructing a bootable CD-ROM for system restore purposes, as this is a common requirement for embedded Linux applications. For instance, if your user needs to replace or upgrade the hard drive, a reinstallation CD is the easiest way to "bless" the blank drive. You could, of course, use the same basic infrastructure to build a "boot application" disk to perform any desired function, in exactly the same way as we prepared the bootable Compact-Flash card above. For instance, you could put a movie player in the image file, and put the movie itself in the main filesystem of the CD.

The principal advantages of CD-ROMs are their read-only nature, and their low cost. These features make the CD ideal for distributing updated software versions and emergency recovery tools to your users. However, you should think carefully about the installation environment before using a bootable CD as the main operating system storage device in a turnkey embedded system. CD-ROM drives are very complex electromechanical systems with free access to the atmosphere, and as such they are horribly sensitive to dust, humidity, vibration and orientation issues. They also have slow access times and a relatively short lifespan. If you decide to use a CD as the primary boot medium for your appliance, then you may wish to consider, at the very least, installing the electronics in an airtight enclosure with vibration-damping mounts. Most CD-ROM drives, especially the slim laptop drives favored by builders of embedded PC systems, do not provide a very good airtight seal when the tray is closed, so you'll have to lock the entire drive away inside a more robust enclosure.

Before we start, we need to decide how to pack up the software restore image we'll be distributing on this disk, a decision which includes the sub-decision "how do we make the target system's hard disk bootable after loading all the files onto it?" The easiest way to do this is simply to start in the root directory of your active system and tar cvfz everything you need—**including the install program for your chosen bootloader**—into one massive image. (The reason for that special instruction will become apparent shortly). I'm going to assume that you have either enough space in your root filesystem for a tarball of everything you need, or that you can mount some other volume temporarily to hold the file; I normally mount a scratch volume over

NFS or SMB. For the purpose of this discussion, I will assume your scratch location is /scratch—if not, modify the following text as appropriate.

Now, let's look at the root directory of a typical Linux system and determine what needs to be copied and what needs special handling:

/bin, /sbin, /dev – You should copy over everything in here.

/mnt – In order for your system to behave normally, you need to ensure that any mountpoint directories in here are created on the target system.

/proc – These are all pseudofiles. You should *not* copy anything in this directory into your install image; strange things will usually happen if you start trying to archive up the /proc pseudofiles. You must, however, ensure that an empty /proc mountpoint is created during the system restore process.

/var – Most of the files in this directory are logs, temporary files and other ephemeral data that doesn't strictly need to be copied to the target system. However, it can be important to have the complete directory hierarchy under here successfully duplicated; some programs may not start up if they can't open logfiles for writing. The easiest way to handle this directory is simply to copy it across.

/tmp – Similarly to /var, this directory contains only volatile information, none of which needs to be saved. However, you must ensure that the directory exists on the target system.

Here is an example shell script that archives up most of the needed information. I have a script similar to this on my development machines; I put it somewhere in the path and call it "master.sh." **NOTE:** Before running this script, ensure that nothing is mounted inside /mnt !

```
#!/bin/sh
echo Duplicating root filesystem
mkdir /scratch/tmproot
cd /scratch/tmproot
cp -a /bin .
cp -a /boot .
cp -a /dev .
cp -a /etc .
cp -a /home .
cp -a /lib .
cp -a /mnt .
mkdir tmp ; chmod 777 tmp
```

```
mkdir proc ; chmod 777 proc
cp -a /root .
cp -a /usr .
cp -a /var .
tar cvfz ../reload.tar.gz *
cd ..
rm -rf tmproot
```

So now we have a complete, compressed copy of the entire filesystem of our target system at /scratch/reload.tar.gz, and it's time to create the boot/root image that will load it onto the hard disk. Preparing the kernel and root filesystem for a bootable CD-ROM is the same process as preparing a flash filesystem boot image; the only real difference is that our startup script will noninteractively perform the installation steps: partition and format the system's hard disk, untar the root image, create any needed directory structures, install the bootloader, eject the CD-ROM, and reboot.

We can use the exact same infrastructure as we created for the CompactFlash card earlier, with only two different files: an additional file called fdisk.script (in the root directory) and a different sysinit script:

```
#!/bin/ash
HOSTNAME=geode
mount -n -t proc /proc /proc
mount -n -o remount,rw /
#swapon -a -e
>/etc/mtab
mount -f /
mount -f /proc
mount -t iso9660 -o ro /dev/hdb /mnt/cdrom
clear
echo "Linux Rescue CD"
echo "==============="
echo ""
echo "ATTENTION!"
echo ""
echo "All files on your system will be erased and the system will be
reloaded"
echo "with a default configuration. If you do not want to do this,"
echo "EJECT THE CD-ROM NOW and SWITCH OFF THE COMPUTER IMMEDIATELY!"
echo ""
echo "The automated restore will continue in fifteen seconds..."
sleep 15
```

```
clear
echo "Restoring System Software"
echo "========================="
echo ""
echo "Setting up hard disk..."
fdisk /dev/hda < /fdisk.script
echo "Creating system area..."
mke2fs -m 0 -j /dev/hda1
echo "Setting up memory..."
mkswap /dev/hda2
echo "Creating user data area..."
mke2fs -m 0 -j /dev/hda3

echo "Copying files to hard disk... (This will take several minutes)"
mount -t ext3 /dev/hda1 /mnt/hda1
cd /mnt/hda1
tar zxvf /mnt/cdrom/reload.tar.gz
umount /mnt/cdrom

echo "Making hard disk bootable..."
chroot /mnt/hda1 /sbin/grub-install /dev/hda
cd /
umount /mnt/hda1
clear
echo "****************************************************************"
echo "*** Restore is complete - System will restart automatically ***"
echo "****************************************************************"
eject /dev/hdb
sync
sleep 2
/sbin/reboot
```

The above script partitions the hard disk, reformats it, extracts the complete system image out of a file called "reload.tar.gz" in the root directory of the CD-ROM, and installs the GRUB bootloader. I've included a complete copy of this reload root-filesystem in the file "cdrom-root.tar.gz" in the root directory of this book's CD-ROM. Note that the fdisk.script file provided in that filesystem partitions the disk into three sections: hda1 (750 MB, ext3), hda2 (128 MB, swap) and hda3 (all remaining space on the disk, ext3). If you want to change this behavior, simply edit the fdisk script, but be aware that you might need to alter the above scriptfile if you change the partition order or on-disk format.

Once you convert the root filesystem directory tree into a compressed image (using the same technique as for the CompactFlash root image; again, call the compressed file initrd.img), you need to put it, along with SYSLINUX and the kernel and syslinux.cfg files (the same files you put onto the CompactFlash card) inside an MS-DOS-formatted 2.88 MB disk image:

```
dd if=/dev/zero of=/288mboot.img bs=1k count=2880
mkfs -t msdos /288mboot.img
syslinux /288mboot.img
mount -o loop /288mboot.img /mnt
cp initrd.img /mnt
cp linux /mnt
cp syslinux.cfg /mnt
umount /mnt
```

That's right—we've buried the root filesystem image inside another disk image! Now all you need to do is fire up your favorite CD burning software, and create a new bootable disk. When prompted for the boot image, select the 288mboot.img file. (Note that this file won't show up in the directory of the CD itself. Also be careful to specify to your CD burning software that the image is a 2.88 MB disk; some software won't autodetect this). Put the reload.tar.gz file, containing your entire backed-up filesystem, into the root directory of the CD. Burn the disk, and voila! You have an automatic system-restore CD.

Final note: I have chosen to use a 2.88 MB El Torito boot-area disk image for our example because it gives us a lot of space for useful programs, and I know by experiment that the BIOS in the PCM-5820 supports it. However, old BIOSes may not support bootable CD-ROMs with "2.88 MB floppy" boot image areas. After some experimentation, I have found that the safest heuristic to use is: when in doubt, make your boot image a 1.44 MB diskette. This is simply a general rule for the broadest possible compatibility; if you are selecting the target hardware on which you intend the software to run, then you can safely ignore this rule.

4.6 Using the Parallel Port as a General-Purpose I/O Interface in Linux

The parallel port is officially a dying interface on the PC, having been largely supplanted (at least, for its original purpose of connecting printers) by USB. However, despite Microsoft's drive for legacy-free systems, the majority of PC-compatibles still have a parallel port. SBC platforms, being largely based on slightly older chipsets and produced on much longer guaranteed delivery schedules, also have parallel ports, and are likely to sport them for many years to come.

Many SBC platforms provide a few bits of general-purpose I/O for application-specific functions. These are often "spare" GPIOs that are provided by multifunction chipsets to support various system functionality that the SBC manufacturer decided not to implement. However, there is no standardized method for accessing these bits, and in many cases there is, incredibly, no documentation available; you need to find and grub through the datasheets for the chipset used on the board in order to work out how to configure and access those I/O lines. There are other issues, too, such as signal voltage conversion, which may bite you (the GPIOs might run at the board's internal I/O voltage, or at 5 V, and if they run at some internal voltage, they might or might not be 5 V-tolerant). All these details are, unfortunately, frequently not well-documented by the board manufacturer. Worst of all, there is rarely any direct support for these proprietary interfaces in third-party operating systems such as Linux. As a result, it's much more portable and generally less involved just to reach for the parallel port when you need a handful of general-purpose I/O bits.

In this section, I'll give you a brief introduction to the PC parallel port hardware and how to use it as a simple pipeline to the outside world. This is by no means a complete reference; it's just enough information for you to be able to interface E2BUS peripherals and understand what the example Linux code I provide is doing. If you require more information, I recommend "Parallel Port Complete: Programming, Interfacing & Using the PC's Parallel Port," by Jan Axelson, ISBN 0-9650-8191-5, Pub Resource 1997. A vast quantity of useful information is also provided freely on the Lakeview Research web site, *http://www.lvr.com/*.

A standard PC/AT architecture supports up to three parallel ports, with base I/O addresses of 0x3BC, 0x378 and 0x278. The port is accessed through three 8-bit registers mapped into the processor's I/O space. These registers are referred to as the data port, the control port and the status port. Standard PC/AT-compatibles name the three parallel ports LPT1, LPT2 and LPT3. These names are **purely historical artifacts of MS-DOS** and have no direct relationship to I/O port assignments. At boot time, the BIOS writes a test byte to the data port of each possible parallel interface, and attempts to read it back. If the readback matches, a parallel interface is assumed to be installed at that address, and a table in the BIOS data area is updated to indicate that fact. A system with three parallel ports installed would have the following I/O port assignments:

Port	Data	Status	Control
LPT1	0x3BC	0x3BD	0x3BE
LPT2	0x378	0x379	0x37A
LPT3	0x278	0x279	0x27A

If one or more of the possible port addresses is not populated, the remaining ports move up to fill the space. For instance, if your system has only one physical printer port, it would be referred to (by DOS and/or Windows) as LPT1 regardless of what port address it's at. Because of this issue, software that accesses parallel port hardware directly uses all sorts of confusingly inconsistent nomenclature. Some software probes the BIOS data area or Windows registry and uses the same names for ports that MS-DOS would use. Some software uses the fixed table above, or a variant of it, to map port names to I/O addresses. (One common variant is to call 0x378 LPT1, 0x278 LPT2, and 0x3BC LPT3—the reason for this is that 0x3BC was originally reserved for parallel ports integrated onto a CGA or MDA graphics card. Since those cards are very rare now, the first—LPT1—port in a PC system is more usually 0x378). Some software requires you to specify the actual port address. It's a mess; fortunately, there are only three possibilities, so trial and error is a reasonable method for getting things to work!

Regardless of this slightly irritating mash of nomenclature, the physical mapping of parallel port pins to I/O register bits is described in the following table (note that some references will give these signals slightly different names):

Pin	Name	Function	Dir	Port	Invert
1	_STROBE	Data strobe	Out	Control bit 0	Yes
2–9	D0-D7	Data bits 0–7	Out	Data bits 0-7	No
10	_ACK	Acknowledge	In	Status bit 3	No
11	_BUSY	Busy	In	Status bit 4	Yes
12	PEND	Paper out	In	Status bit 2	No
13	SELECT	Device select feedback	In	Status bit 1	No
14	_AUTOFD	Auto linefeed	Out	Control bit 1	Yes
15	_ERROR	Error	In	Status bit 0	No
16	_INIT	Initialize (reset)	Out	Control bit 2	No
17	SELECTIN	Device select	Out	Control bit 3	Yes
18–25	GND	Ground	–	–	–

There are several types and operational modes of modern parallel ports. In the lowest common denominator configuration, the status register is read-only, the control register is write-only, and the data register can be considered as write-only also. (This isn't strictly true, of course—for example, if it *was* true, the BIOS's autoprobing technique, which I outlined briefly earlier, wouldn't be possible—but for our purposes it's an acceptably approximate description of the port's behavior. If you want more details on how the parallel port is really constructed, consult one of the references I mentioned earlier). Practically all modern PC systems support several modes of operation selectable in CMOS set. These modes are usually SPP, EPP1.7, EPP1.9 and ECP. Although theoretically our circuit should work with any of these modes configured, in practice you are likely to encounter problems if your PC is configured for ECP. I suggest that you configure the port for SPP mode unless you have some special reason for needing a different mode.

One final note here, before we go on to software issues: You should be aware that there are various bizarre electrical issues with certain PC-compatibles, particularly older name-brand machines, and especially portable machines. The only thing that you can say for certain about a "standards-compliant PC" parallel port is that for some specified drive current, it defines the output "high" voltage as 2.4 V, and the output "low" voltage as 0.4 V, with anything in between being illegal. This leaves a lot of room for different implementations to bite your design. In general, if you're designing a device that has to have a good chance of working on a wide range of PCs:

- Don't make heavy drive demands on any output lines. Three or four TTL loads is probably a safe cutoff point. Particularly if you plan to run any significant length of cable between the PC and your circuit, it's best to run the lines straight to a buffer in your device, and fan out from there.

- Never connect parallel port output lines directly to any logic that isn't 5 V-tolerant at your chosen internal supply voltage. Many (if not all) modern motherboards and portable machines will show a maximum voltage swing of 0 V to +3.3 V (reflecting the V_{IO} of 3.3 V found in modern motherboard logic), but you certainly can't rely on this.

- It's advisable to have Schmitt triggers between your circuit's inputs and the parallel port output lines you intend to use. Some of the control lines are implemented as an open-collector driver, with a pullup resistor and a capacitor to ground to limit the risetime.

With older PCs, it was usually (but not certainly not always) possible to avoid many of these problems by going into the BIOS setup menu and configuring the parallel port for SPP mode. Unfortunately, many modern PCs have eliminated even that last mousehole—they are locked into ECP mode, and it can be difficult or impossible to get them to work with some parallel-connected peripherals. (The factory-supplied ICEs and flash programmer hardware for several microcontrollers fall into this category of ill-behaved appliances, by the way).

Now that we know how the parallel port works, how do we access it inside Linux? There are a several possibilities here. First, we could write a kernel driver to do what we need to do. The kernel driver would have full I/O privileges and could write to the parallel port I/O addresses directly. Second, we could jump through a couple of hoops and write a user-mode program that poked directly to the parallel I/O addresses. Finally, we could use the standard Linux parallel port device API, ppdev. I recommend this latter method unless there are really compelling reasons to choose otherwise. Writing a kernel driver is just a distraction—it certainly gives you fairly complete control over the process, including allowing you to disable interrupts while you twiddle the port bits, but it's frankly overkill for the kind of thing we're doing. Writing to the port addresses directly from user mode is possible, but it's clumsy. We would also have to write some code to work out where the parallel ports are installed in the machine's I/O space. Using the standard API avoids a lot of unnecessary tweaking and it adds very little overhead—this API was designed for use by drivers like the parallel port ZIP device, so it's capable of quite high throughput.

To use the parallel port, we open a file descriptor to /dev/parport*n* (where *n* is the parallel port number—we're going to use parport0, i.e., LPT1, for all our sample code) and use ioctl on this descriptor to communicate with the underlying driver. If you are testing this code with anything other than the kernel configuration I have supplied, note that you MUST disable parallel *printer* support—but not parallel *port* support—in the kernel for the code in this book to work.

The first thing to do is open the parallel port device and acquire exclusive control over it. This is simply achieved with something similar to the following code fragment:

```
#include <stdio.h>
#include <fcntl.h>
#include <linux/parport.h>
#include <linux/ppdev.h>
#include <sys/ioctl.h>

int handle = 0;

handle = open("/dev/parport0", O_RDWR);
if (handle < 0) {
    // report fatal error
}
// Get exclusive access to port
ioctl(handle, PPEXCL, 0);
ioctl(handle, PPCLAIM, 0);
```

Don't throw that handle away; we'll use it for accessing the other functions described next. Now, how to wiggle the pins on the port? The parallel port control register is accessed through the PPFCONTROL ioctl. This call takes a pointer to a ppdev_frob_struct structure consisting of a mask byte and a new data byte. This way, you can toggle individual control bits—potentially, different bits from multiple threads—without having to maintain your own copy of the current port state. This structure, found in <linux/ppdev.h>, is defined as follows:

```
struct ppdev_frob_struct {
    unsigned char mask;
    unsigned char val;
};
```

To set or clear bits in the control register, set the bits in `mask` corresponding to the bits that you wish to change, and load the desired new data into `val`. Then simply call ioctl(handle, PPFCONTROL, &fs), where fs is the name of your ppdev_frob_struct. To make life easy for you, the ppdev.h header contains definitions for the control bits:

```
#define PARPORT_CONTROL_STROBE    0x1
#define PARPORT_CONTROL_AUTOFD    0x2
#define PARPORT_CONTROL_INIT      0x4
#define PARPORT_CONTROL_SELECT    0x8
```

For example, to clear the STROBE bit, you would use the following code:

```
fs.mask = PARPORT_CONTROL_STROBE;
fs.val = 0;
ioctl(handle, PPFCONTROL, &fs);
```

Reading the status register is accomplished using the PPRSTATUS ioctl. Again, the kerne includes define names for the bits:

```
#define PARPORT_STATUS_ERROR      0x8
#define PARPORT_STATUS_SELECT     0x10
#define PARPORT_STATUS_PAPEROUT   0x20
#define PARPORT_STATUS_ACK        0x40
#define PARPORT_STATUS_BUSY       0x80
```

You use the PPRSTATUS ioctl simply by passing a pointer to a 1-byte buffer to store the port's current status:

```
unsigned char c;
ioctl(handle, PPRSTATUS, &c);
```

Reading and writing the data lines is achieved using the same syntax as for PPRSTATUS, but the two ioctls are PPRDATA and PPWDATA, respectively. Note that before reading or writing the data bus, you must ensure that it has been set into the correct mode (input or output) using the PPDATADIR ioctl. For example, take this code, which reads a byte from the data bus, then writes a different byte:

```
int i;
char c1, c2;

// An arbitrary byte to be written to the port
c2 = 0xf5;

// Set port to input mode
c1 = 0xff;
ioctl(handle, PPWDATA, &c1);
i = -1;
ioctl(handle, PPDATADIR, &i);

// Read byte from port
ioctl(handle, PPRDATA, &c1);

// Write new data to port and set it to output mode
ioctl(handle, PPWDATA, &c2);
i = 0;
ioctl(handle, PPDATADIR, &i);
```

Note how we write 0xFF to the output latch before we read the port. The reason for this is that when the port is in readback mode, the output latches are open-collector. Writing 0xFF to the data latch effectively tristates the outputs so they don't try to pull down signals from the outside world. Also note that you can't really mix and match—either the entire port is an input, or the entire port is an output. You can't specify the data direction at a finer resolution, although there are terrifying hardware tricks to work around this limitation.

4.7 Implementing Graphical Control Interfaces

4.7.1 Introduction

In this section, I'm going to give you a very short overview of the options available to you when implementing graphical control/overmonitoring interfaces on Linux systems. In particular, I'm going to concentrate on interfaces that lend themselves well to embedded-friendly feature paring. This text is obviously not intended to be an in-depth how-to guide for any of the specific graphics systems I mention. It's intended to show you the advantages and disadvantages of a number of different possible GUI choices, and provide you with some pointers to further research into the options you like the best.

The issue of implementing graphics functionality on your system really breaks down into two subproblems: how you get the system into a correctly initialized graphics mode, and what you have to do in order to get graphical elements onto the screen once the graphics mode is set.

The first thing I'd like to impress on you is the horrible but virtually irresistable temptation to re-invent wheels. Many projects that need or want a graphical interface start out with extremely modest needs; for example, some simple bitmapped graphics and a single text font. For such a tiny amount of code, it seems that the most efficient approach is to write your own graphics routines entirely from scratch rather than invest a lot of time climbing the learning curve for an off-the-shelf library. There are two hidden flaws in this piece of logic: first, the golden rule is that *all* projects mushroom beyond their initial idea (meaning that one day you'll inevitably find yourself slaving to implement and debug a hand-rolled version of some tricky function that you could have just called out of a pre-existing library), and second, you'll probably have to repeat a lot of this pedestrian work every time you start a new project. Both these issues are more or less avoided, and your life is made much simpler, if you pick a reasonably portable graphics library and use it across multiple projects.

Essentially, there are only a handful of good reasons to roll your own (and even these reasons are probably arguable):

- You are doing something truly unique and fundamentally incompatible with the design paradigm of any extant graphical library. I have only once been involved in such a project[26]: designing a GUI based around hexagons instead of rectangles. All screen surfaces, windows, gadgets, etc. were expressed in terms of six side-lengths instead of the normal width and height parameters in a rectangular coordinate system.

- Your target hardware has some special acceleration features or other hardware magic for which you would need to write drivers anyway. Perhaps you can gain better performance by designing your GUI's structure around the capabilities of the accelerator hardware, rather than writing a driver for an existing GUI.

- You have unusually strict performance requirements—real-time issues, memory consumption, and so on. An example of the type of system that satisfies this condition (as well as the previous condition, usually) is a low-end digital camera. These devices are often based on 8051-cored ASSP with specialized JPEG compression hardware on-chip.

- You need to maintain rigorous control over the portability and platform-independence of the code. For instance, you might need to support two or three different hardware platforms (and no others), and you might want to make design optimizations specific to those particular platforms.

- You need to be able to test, certify and guarantee every line of code that is going into the final system, for security or reliability purposes (or some similarly critical reason). Writing a proprietary GUI can save you an enormous amount of work on the back end of the project if you have a requirement like this. Imagine how many man-hours would be required to perform a complete sourcecode audit on, say, XFree86!

---

[26] It was a very silly project, too—the client wanted a user interface in Klingon. Avid watchers of Star Trek® will note that most of the Klingon computer displays use hexagonal grids and controls. Don't expect projects like this to come along every day.

Having done my best to steer you towards the straight and narrow, let's consider some of the choices open to you. For each option, I'll provide some minimal sample code so you can get a simple application of your own up and running. Where appropriate, I'll also point you to some further reading on the topic.

4.7.2 The Framebuffer Graphics Console (fbdev)

The framebuffer graphics console is often used in embedded systems, particularly systems that are based around non-x86 microcontrollers with built-in display controller hardware. If your particular hardware combination is supported by the kernel, it provides a simple way to get the system into a graphics mode and to query the address of video RAM. You can then either use your own proprietary GUI code, or port one of the many graphics libraries (such as Qt-embedded) to implement the actual interface portion of the code. If you are using x86 Linux to prototype something that will eventually be squeezed into an ARM, MIPS or similar SOC (system-on-chip) device, the framebuffer driver is almost certainly your line of least resistance.

Geode's graphics hardware (CS5530) isn't explicitly supported by a Linux framebuffer driver, but there is a generic driver that uses VESA BIOS Extension (VBE) calls to set video modes, which you can use to prod the system into a graphics mode at boot time. The sample kernel configuration I included on the CD-ROM includes the VESA graphical framebuffer driver. If you reboot your SBC, hit any key at the GRUB prompt, and edit the kernel boot line to include the command "vga=0x311," your system will start up in a 640 × 480, 16 bpp graphics mode. (For more information on VGA mode numbers, refer to Documentation/fb/vesafb.txt in your Linux kernel source directory).

Here's a basic outline of how to use framebuffer mode: First, make sure that the system is in a graphics mode by editing your kernel command line as I just described, and rebooting the system if necessary. Next, open a handle to the framebuffer device of interest (probably /dev/fb0). Use the FBIOGET_FSCREENINFO and FBIOGET_VSCREENINFO to obtain fixed and variable (mode-specific) screen information, respectively. This information is necessary to calculate the row stride (bytes per scanline), determine the size of the framebuffer memory to be mapped into your process's address space, and otherwise locate bytes onscreen.

Note that if you're reading this section, you'll most likely have selected a specific resolution and bit depth for your application in advance, and your code will be targeted specifically to that bit depth. Your screen layouts will probably also be tailored specifically for a certain screen resolution. For this reason, in embedded applications you'll most likely simply be validating the system settings against your hardcoded constraints, rather than inspecting the system and dynamically adapting your code to match the hardware's capabilities.

Here is some illustrative code for you (this program is included in the fbtest directory of the sample programs archive):

```c
/*
   main.c

   Demonstration applet for framebuffer

   From "A Cookbook for Open-Source Robotics and Process Control"
   Lewin A.R.W. Edwards (sysadm@zws.com)
*/

#include <stdio.h>
#include <fcntl.h>
#include <string.h>
#include <sys/ioctl.h>
#include <sys/mman.h>
#include <linux/fb.h>

#define FB_DEVICE "/dev/fb0"

int main (int _argc, char *_argv[])
{
    int handle,i,j,screensize;
    unsigned char *framebuffer,*backup;
    struct fb_fix_screeninfo fi;
    struct fb_var_screeninfo vi;

    // Open framebuffer device
    handle = open(FB_DEVICE, O_RDWR);
    if (handle == -1) {
        printf("Error opening " FB_DEVICE ".\n");
        return -1;
    }
```

```
   // Get fixed screen information and show an informative message
   ioctl(handle, FBIOGET_FSCREENINFO, &fi);
   printf("Device is '%s'.\n", fi.id);
   printf("Buffer: 0x%-08.8X bytes at physical address 0x%-08.8X.\
nMMIO at 0x%-08.8X, accel flags %-08.8X.\n", fi.smem_len, fi.smem_start,
fi.mmio_len, fi.accel);
   printf("%d bytes per physical scanline.\n", fi.line_length);

   // Get variable screen information and show an informative message
   ioctl(handle, FBIOGET_VSCREENINFO, &vi);
   printf("Currently viewing (%d,%d) window of (%d,%d) display
at %dbpp.\n", vi.xres,vi.yres, vi.xres_virtual, vi.yres_virtual,
vi.bits_per_pixel);


   screensize = vi.xres_virtual * vi.yres_virtual * (vi.bits_per_pixel
/ 8);

   framebuffer = mmap(0, screensize, PROT_READ | PROT_WRITE, MAP_
SHARED, handle, 0);
   if (!framebuffer) {
      printf("Error mapping framebuffer into process.\n");
      return -1;
   }



   // Allocate memory for backup screen and copy it
   backup = malloc(screensize);
   if (!backup) {
      printf("Cannot allocate memory for framebuffer backup.\n");
      return -1;
   }
   memcpy(backup, framebuffer, screensize);

   // Wait for a few seconds, then show some coruscating colors
   sleep(3);
   for (i=0;i<256;i++) {
      for (j=0;j<640*480*2;j++)
         framebuffer[j]=i;
   }

   // Restore original screen contents
   memcpy(framebuffer, backup, screensize);
   return 0;
}
```

The main disadvantages to using the framebuffer are:

1. You generally have little or no access to hardware acceleration features. There are a handful of platform-specific framebuffer drivers in current kernels, and these support a small core of acceleration functionality, but the generic VESA driver is unaccelerated. Working with the framebuffer is a bit like asking for a salad, and being handed a shovel and a packet of seeds; it's possible to do almost anything, but it can be time-consuming.

2. It's often not possible to change to any different video mode at runtime—you boot up in a certain mode, and you're stuck there. Although there are APIs in the framebuffer code to change video modes, this part of the code is quite unstable (again, particularly the VESA-based code).

3. In the case of the VBE-based framebuffer device, there may be BIOS issues that make the graphics modes quirky. Unfortunately, this fact is true for all versions of Geode BIOSes that I have tested.

The framebuffer device does, however, have the advantage of being very simple to work with. Another useful characteristic of the framebuffer code you'll be writing is that it's highly portable to other systems—including OS-less systems—as long as they have the same pixel format. Because there are almost no required APIs except the startup job of ascertaining the graphics mode and start location of video memory, using fbdev doesn't tie you down by forcing you to make an investment in any particular software architecture. When Linux is ported to new platforms (microcontroller evaluation boards, PDAs, video game consoles, set-top boxes and so forth), invariably the first graphics subsystem ported is the framebuffer console, so if you work with the framebuffer you'll always be able to explore the leading edge of new hardware ports.

4.7.3 SVGAlib

SVGAlib is a couple of steps more advanced than the dumb framebuffer. It includes APIs for a few useful functions, and on some platforms can take advantage of hardware acceleration features. It was originally designed for games and emulators, which means that it supports handy animation-friendly features such as double-buffering. (In fact, SVGAlib is an evolutionary phenomenon based on an older, less general-purpose library called vgalib).

The current release version of svgalib (1.4.3) is very slightly syntactically incompatible with current versions of gcc, so it isn't possible to compile and install this version directly on modern Linux distributions such as Fedora Core 1. To save you some head-scratching time, I have patched the affected file, src/vga.c, and rearchived the tarball, which you will find as linux/svgalib-1.4.3-patched.tar.gz on the CD-ROM with this book. To build and install, simply unpack the tarball, `cd svgalib-1.4.3`, and make install ; make `demoprogs`. After compiling and installation are complete, add the line `/usr/local/lib` to /etc/ld.so.conf (if it isn't there already), and run ldconfig(8) so that SVGAlib programs can find the shared libraries. You'll also want to edit the configuration file /etc/vga/libvga.config to reflect your hardware configuration (at least the video card selection, and possibly also the mouse type). If you're configuring for the Geode SBC family we've been discussing, at minimum you'll need to add the line `chipset VESA` to use the generic VESA BIOS code.

With the library installed and configured, we can start writing some actual code. We're going to write an application that incorporates some of the machine vision code in Section 4.9.1 to capture images from an attached video camera and display them onscreen, with an overlay showing the outlines of sharply-defined objects in the image. The easiest way to show you how to do this is to present the main() meat of the program and go through it line by line, so here it is (this is a listing of the main.c file):

```
/*
    Example svgalib + V4L application - Displays camera input onscreen
    2004-04-03 larwe created
*/

#include <sys/types.h>
#include <sys/stat.h>
#include <stdio.h>
#include <stdlib.h>
#include <fcntl.h>
#include <unistd.h>
#include <sys/ioctl.h>
#include <linux/videodev.h>

#include <vga.h>
#include <vgagl.h>
```

```c
#include "v4lcap.h"

// Key code definitions
#define KEY_ESCAPE    27    // Esc

// Miscellaneous default settings
#define VIDEO_MODE G640x480x64K
#define DEFAULT_NOISEFLOOR      8

// Internal variables
GraphicsContext *phys_screen, *virt_screen;
BMINFO edge_image;
unsigned char noisefloor = DEFAULT_NOISEFLOOR;

/*
   Demonstration main function
*/
int main(int _argc, char *_argv[])
{
    int fQuit = 0, c, blit_x, blit_y;
    int k;

    // Initalize capture device
    if (V4LC_Init(320,240)) {
        printf("Error initializing video capture device.\n");
        return -1;
    }

    // Create a second bitmap structure to hold the derived edge data
    memcpy(&edge_image, &V4L_bitmap, sizeof(edge_image));
    edge_image.bitmapdata = malloc(edge_image.allocsize);
    if (edge_image.bitmapdata == NULL) {
        printf("Could not allocate memory for edge image.\n");
        return -1;
    }

    // Initialize SVGA graphics, physical and offscreen contexts
    vga_init();
    vga_setmode(VIDEO_MODE);
    gl_setcontextvga(VIDEO_MODE);
    phys_screen = gl_allocatecontext();
    gl_getcontext(phys_screen);
```

```
gl_setcontextvgavirtual(VIDEO_MODE);
virt_screen = gl_allocatecontext();
gl_getcontext(virt_screen);

// Select offscreen drawing environment as target for SVGAlib ops
gl_setcontext(virt_screen);

// Select default 8x8 text font
gl_setfont(8, 8, gl_font8x8);
gl_setwritemode(WRITEMODE_OVERWRITE | FONT_COMPRESSED);

// Calculate desired blit size and draw a border around the target area
blit_x = V4L_bitmap.width;
blit_y = V4L_bitmap.height;
if (blit_x > 352)
    blit_x = 352;
if (blit_y > 288)
    blit_y = 288;
gl_line(0, 0, blit_x + 1, 0, 0xffff);
gl_line(blit_x + 1, 0, blit_x + 1, blit_y + 1, 0xffff);
gl_line(0, blit_y + 1, blit_x + 1, blit_y + 1, 0xffff);
gl_line(0, 0, 0, blit_y + 1, 0xffff);

while (!fQuit) {
    int i,j;
    unsigned short pixel, *dest;
    unsigned char *src;
    unsigned short r, g, b;
    char tmps[80];


    // Acquire one frame from the capture device
    V4LC_Acquire();

    // Copy frame data to temp processing area and run edge detection
    memcpy(edge_image.bitmapdata, V4L_bitmap.bitmapdata,
        edge_image.allocsize);
    for (i=0; i<blit_y; i++) {
        DER_ScanlineToGrayscale(edge_image.bitmapdata +
            (i * edge_image.width * 3), edge_image.width);
        DER_DeriveScanline(edge_image.bitmapdata +
            (i * edge_image.width * 3), edge_image.width);
    }
```

```
// Run overlay pass over captured image
for (i=0; i<blit_y; i++) {
   unsigned char *ovldest = V4L_bitmap.bitmapdata +
      (i * V4L_bitmap.width * 3) + 1;

   src = edge_image.bitmapdata + (i * edge_image.width * 3)
      + 1;
   for (j=0; j< blit_x ; j++) {

      if (*src > noisefloor)
         *ovldest = 255;
      src += 3;
      ovldest += 3;
   }
}

// Copy image to the frame buffer (converting to 5:6:5 RGB)
for (i=0; i<blit_y; i++) {
   src = V4L_bitmap.bitmapdata + (i * V4L_bitmap.width * 3);
   dest = (unsigned short *) ((((unsigned char *) VBUF) +
      (i + 1) * BYTEWIDTH) + 2);
   for (j=0; j<blit_x; j++) {
      r = *(src++) & 0xf8;
      g = *(src++) & 0xfc;
      b = *(src++) & 0xf8;
      pixel  = (r << 8) | (g << 3) | (b >> 3);
      *(dest++) = pixel;
   }
}

// Display current processor settings
gl_setfontcolors(0x0000, 0x07e0);
sprintf(tmps,"CapSize    : (%d,%d)         ", blit_x, blit_y);
gl_write(blit_x + 4, 0, tmps);
sprintf(tmps,"NoiseFloor: %d     ", noisefloor);
gl_write(blit_x + 4, 8, tmps);

gl_setfontcolors(0x0000, 0xffe0);
gl_write(0, blit_y + 4,
   "Esc - Exit  W / X - Increment/decrement noise floor");

// Copy current [offscreen] context to visible framebuffer
gl_copyscreen(phys_screen);
```

```
    // Check if there is a keystroke - if so, act on it
    k = vga_getkey();

    switch(k) {
        case 'w':
        case 'W':
            noisefloor++;
            if (noisefloor == 0)
                noisefloor = 255;
        break;

        case 'x':
        case 'X':
            noisefloor--;
            if (noisefloor == 255)
                noisefloor = 0;
        break;

        case KEY_ESCAPE:
            fQuit = -1;
        break;
        default:;
    }
}

// Clean up display environment
gl_clearscreen(0);
vga_setmode(TEXT);
}
```

The first few lines are housekeeping setup for the video capture functions and a few data structures that are required by the image-processing routines. The first svgalib-specific call is vga_init(). This function should be called very early in your program (certainly before using any other svgalib functions); it installs handlers to catch various signals so that if your program dies, svgalib gets a chance to restore the video environment.

Next, we set an extended video mode with vga_setmode(). The mode constants are defined in vga.h with easy-to-understand mnemonic names. (Note that the "x" character in the mode names is always lowercase; I invariably mistype this when writing svgalib code). Changing video modes is probably the most stressful thing svgalib will do to your system. The potential dangers vary according to the type of chipset support you're using, and what your definition of a fatal problem might be. In a worst-case scenario, mode changes can lock up the system so hard that a power-cycle or hard reset (or a watchdog bite, if your system is equipped with suitable WDT hardware) is the only way to resurrect it. This scenario is, unfortunately, most likely to occur when you're using VESA BIOS support—VESA BIOS extensions and ACPI tables seem to be the most bug-prone and haphazardly tested pieces of code in a modern PC. Note that if you happen to be running your svgalib program within XFree86, svgalib will automatically allocate a new console and switch to it—the XFree86 video state will be preserved and in theory it should be possible to return to X after your program terminates. In practice, this facility is horribly broken on a lot of platforms, including Geode[27]. The soundest advice I can offer you here is that if you're using svgalib, you should set a single video mode and never change it—and if you're using XFree86, don't mix in svgalib applications.

Having presumably been successful in pulling the system into our chosen video mode (640 × 480, 5:6:5 16 bpp direct color), we now use gl_vgacontextvga( ) to set the current graphics context to reflect the physical screen's parameters, allocate a new context variable with gl_allocatecontext( ) to store these parameters, and copy the current graphics context into this newly allocated variable with gl_getcontext( ). Similarly, we allocate an offscreen drawing surface with the same characteristics as the physical screen, and we point the SVGAlib functions to work on this offscreen buffer with gl_setcontext( ). The next few lines draw the screen layout and onscreen help. Remember that these items are all being drawn into the offscreen buffer; they're invisible for the moment.

We now begin the main program loop, which loops continuously until the quit flag is set by a keystroke. The V4L_Acquire( ) function grabs a single frame from the webcam (refer to the sourcecode on disk for the gory details of this process). We then process it a little bit according to the current settings, superimpose the edge overlay, and copy it into the offscreen buffer. After the complete frame is assembled, we use gl_copyscreen( ) to blit the entire offscreen buffer onto the active display area. Fi-

---

[27] Like most comments about Geode compatibility, this issue depends on your BIOS version.

nally, we use the vga_getkey() function to get a keystroke, if any, out of the keyboard buffer, and if the user has pressed a key we process it appropriately.

All of this gymnastic activity sounds like a lot of manipulation, but even the Geode manages to achieve a fairly respectable framerate (about 15 fps).

If you remember the dark ages of programming graphics in DOS using the graphical console libraries provided by Watcom (now Sybase), Borland and Microsoft, you won't have much trouble working with svgalib.

4.7.4 X

With its minimalist name but a far-from-minimalist architecture, X is the *non plus ultra* of graphical interfaces. Anything you want to do can be done in X—the question is merely if you'll still want to do it once you find out how much work it's going to be. The old tagline says "Programming graphics in X is like computing sqrt(pi) in Roman numerals," and with good reason. If you intend to use X as your environment, you will probably be using a wrapper library to make your life easier, but it's still not a trivial matter to develop an X application. If I can extend the analogy I made in my discussion of the framebuffer (making a salad from seeds), developing X applications is often like asking for a salad and being given a printout of the lettuce and carrot genomes; technically, all the information you need is there, but sometimes it seems as if alien technologies are required to synthesize the desired product from the available ingredients.

X itself is also terribly resource-intensive and the architecture has inherent performance bottlenecks. There are various extension features, with differing degrees of portability, to alleviate these bottlenecks, but they aren't universally available and add yet more complexity to your program. My recommendation is to eschew X in embedded systems unless you need to be able to run extant third-party programs that require it, or you already have an X-based desktop application that you are pruning down for embedded use. X was designed to solve a variety of technical problems that simply don't exist in the majority of embedded systems; primarily, it was designed to provide a GUI layer over a communications session with applications running on a remote machine. This introduces all kinds of irritating assumptions and schizophrenic bottlenecks due to the fact that the GUI and the programs running inside it are conceptually on opposite ends of a network connection. Shared memory techniques are only a workaround, not a solution—they don't address the conceptual limitation that certain resources and data structures are "server side" and certain structures are "client side."

Unfortunately, the best way of escaping the program complexity issues (besides using some other GUI) is to use a highly abstracted programming environment like Java. Of course, that approach introduces its own frustrations. I wouldn't advise attempting to write any complex Java programs for use on Geode; although they do run at a semi-reasonable speed with JIT compilation, there can be literally minutes of startup delay for even relatively simple programs.

If you do want to use X on the Geode platform, you should be aware that there are some hoops to jump through. I've assembled here some special notes that may help you to get it working properly for your situation. There are three major routes you can take towards bringing up the X server:

Method 1 – Use the framebuffer console and the XF86_FBDev server. This method might not work on all Geode-based systems because it relies on the presence of a VESA BIOS extension. It works on the PCM-5820 (but note the following important drawbacks). For testing purposes, you can simply type "video=vesa vga=xxx" at the LILO boot: prompt, or make whatever changes are appropriate to your boot-loader if you're not using LILO. This will allow you to check various video modes (xxx = video mode; look in Documentation/vb/vesafb.txt inside your Linux kernel source directory for more information on this). Once you've established which mode works best for you, adding these lines to the appropriate paragraph in lilo.conf, or modifying your bootloader's will make your choice permanent:

```
vga=xxx
append="video=vesa"
```

Note that this feature requires kernel framebuffer console support. Assuming you have everything working for the framebuffer console, simply install the XF86_FBDev server and link it to /usr/X11R6/bin/X and you'll be set. Following is an example XF86Config file for framebuffer operation. Using this file, X will start up in the resolution you selected at boot time.

```
Section "Files"
RgbPath "/usr/X11R6/lib/X11/rgb"
FontPath "unix/:7100"
EndSection

Section "ServerFlags"
EndSection

Section "Keyboard"
Protocol "Standard"
AutoRepeat 500 5
LeftAlt Meta
RightAlt Meta
ScrollLock Compose
RightCtl Control
XkbKeycodes "xfree86"
XkbTypes "default"
XkbCompat "default"
XkbSymbols "us(pc101)"
XkbGeometry "pc"
XkbRules "xfree86"
XkbModel "pc101"
XkbLayout "us"
EndSection

Section "Pointer"
Protocol "PS/2"
Device "/dev/mouse"
Emulate3Buttons
Emulate3Timeout 50
EndSection

Section "Monitor"
Identifier "Panel"
VendorName "Unknown"
ModelName "Unknown"
HorizSync 31-90
VertRefresh 40-160
# Modelines aren't actually used
Modeline "640x480" 31.5 640 656 720 864 480 488 491 521
EndSection

Section "Device"
Identifier "Geode"
VendorName "Unknown"
```

```
BoardName "Unknown"
VideoRam 4096
EndSection
Section "Screen"
Driver "FBDev"
Device "Geode"
Monitor "Panel"
Subsection "Display"
Depth 16
Modes "default"
ViewPort 0 0
EndSubsection
EndSection
```

There are two major disadvantages to this setup (and some minor ones). The two big problems are (a) it's slow, because you have no access to any accelerated hardware features, and (b) you can't switch resolutions on the fly. In the multimedia application I mentioned in the introduction to this book, we wanted to run at 1024 × 768, 16 bpp for still images, but use lower resolutions for full-screen motion video (MPEG-1 playback) since the source material is low resolution anyway, and stretching it out to fill a large screen is a complete waste of CPU bandwidth.

Important note: If you are using the VESA framebuffer console, it is not advisable to use non-FBDev X servers, nor graphics libraries like SVGAlib. The system will often go into an undefined video mode as soon as you attempt to switch modes away from the boot-time default.

If those disadvantages put you off Method 1, consider **Method 2** – Use the XF86_ SVGA server from XFree86 version 3.x. The best match I have found so far is the XF86_SVGA server, version 3.3.6a. If you install Red Hat 7.2 and select Geode, this server will be installed but it will *NOT* be enabled! The Red Hat 7.2 install process by default links /etc/X11R6/bin/X to /etc/X11R6/bin/XFree86, which is the XFree86 4.1.0 "mega-wrapper"; it attempts to install the Cyrix MediaGX driver, which does not work with Geode. Newer Linux distributions don't usually include 3.x servers at all.

Note that the 3.3.6a server I mention here is NOT the same version Advantech supplies on their driver CD-ROM. The Advantech version seems to calculate dot clocks and other timings quite bizarrely, and I've had trouble getting properly centered video with their server. Unfortunately, there's a catch-22 lurking here, which I describe next.

One essential point to note is that there is a bug in the CS5530, or more likely the VSA code (yet again!) which can make the SVGA server extremely unstable. You can work around this bug by specifying an odd virtual screen size using the Virtual x y keyword (I use 1024 × 769 because our system normally runs at 1024 × 768. If you are using a lower screen resolution like 640 × 480 and want to conserve video memory requirements, use an appropriately odd-sized virtual buffer such as 640 × 481). The cause of this bug is a combination of XF86_SVGA's behavior and a quirk of the CS5530. The Geode system relies heavily on I/O traps and "faked" hardware emulation for some of its functionality, particularly video. In the "standard" resolutions, XF86_SVGA attempts to enable display compression, which causes problems. In the best case, you will get some garbage on the display; in the worst case, the system will lock hard and require a hard reset or power-cycle. Specifying the strange virtual resolution implicitly disables the compressed display feature and works around the problem. This workaround may or may not be ideal for you depending on your system setup; in the case of the application I was implementing, the system has no pointing device so it's not possible for the user to scroll the display window and see the extra "phantom" scanline.

Using XF86_SVGA you will have access to some hardware acceleration features such as a hardware mouse cursor. More importantly, you will be able to change resolutions on the fly. Here is a suitable XF86Config file for the three standard resolutions, 640 × 480, 800 × 600 and 1024 × 768. It is possible to get the low-resolution video modes working also, but only with the Advantech-supplied server. (Argh! This is the catch-22 I mentioned.) If you don't need low resolution capabilities, I suggest you stick to the Red Hat-supplied server. It's much easier to get stable, easily-centered video with that server.

```
Section "Files"
RgbPath "/usr/X11R6/lib/X11/rgb"
FontPath "unix/:7100"
EndSection
```

```
Section "ServerFlags"
EndSection

Section "Keyboard"
Protocol "Standard"
AutoRepeat 500 5
LeftAlt Meta
RightAlt Meta
ScrollLock Compose
RightCtl Control
XkbKeycodes "xfree86"
XkbTypes "default"
XkbCompat "default"
XkbSymbols "us(pc101)"
XkbGeometry "pc"
XkbRules "xfree86"
XkbModel "pc101"
XkbLayout "us"
EndSection

Section "Pointer"
Protocol "PS/2"
Device "/dev/mouse"
Emulate3Buttons
Emulate3Timeout 50
EndSection

Section "Monitor"
Identifier "Panel"
VendorName "Unknown"
ModelName "Unknown"
HorizSync 31-90
VertRefresh 40-160
# 640x480 @ 72 Hz, 36.5 kHz hsync
Modeline "640x480" 31.5 640 656 720 864 480 488 491 521
# 800x600 @ 72 Hz, 48.0 kHz hsync
Modeline "800x600" 50 800 816 976 1040 600 637 643 666 +hsync
+vsync
# 1024x768 @ 70 Hz, 56.5 kHz hsync
Modeline "1024x768" 75 1024 1040 1184 1328 768 771 777 806 -hsync
-vsync
EndSection

Section "Device"
Identifier "Geode"
```

```
VendorName "Unknown"
BoardName "Unknown"
VideoRam 4096
EndSection

Section "Screen"
Driver "svga"
Device "Geode"
Monitor "Panel"
DefaultColorDepth 16
Subsection "Display"
Depth 16
Modes "1024x768" "800x600" "640x480"
Virtual 1024 769
EndSubsection
EndSection
```

One final note on XFree86 3.x—National Semiconductor has supplied numerous subtly different versions of the 3.x SVGA server for Geode; some with sourcecode, some as binary-only. Since 3.x is officially dead, I have not experimented with all of these. If you're desperate to fix some particular issue, feel free to go the trial-and-error route!

If neither Methods 1 nor 2 appeal to you, try **Method 3** – Use XFree86 4.x. Modern distributions of Linux, such as Red Hat 9.0 and Fedora Core 1, ship with XFree86 4.x. However, the installation process doesn't correctly detect the CS 5530 chipset; it installs the Cyrix MediaGX driver, again. To get the system working properly, use this XF86Config, which uses the nsc_drv.o driver. Note that this XF86Config has only been tested on XFree86 4.2.99.3 beta and the current release, 4.3.0.

```
Section "ServerLayout"
Identifier "XFree86 Configured"
Screen 0 "Screen0" 0 0
InputDevice "Mouse0" "CorePointer"
InputDevice "Keyboard0" "CoreKeyboard"
EndSection

Section "Files"
RgbPath "/usr/X11R6/lib/X11/rgb"
ModulePath "/usr/X11R6/lib/modules"
FontPath "/usr/X11R6/lib/X11/fonts/misc/"
```

```
FontPath "/usr/X11R6/lib/X11/fonts/Speedo/"
FontPath "/usr/X11R6/lib/X11/fonts/Type1/"
FontPath "/usr/X11R6/lib/X11/fonts/CID/"
FontPath "/usr/X11R6/lib/X11/fonts/75dpi/"
FontPath "/usr/X11R6/lib/X11/fonts/100dpi/"
EndSection

Section "Module"
Load "extmod"
Load "dbe"
Load "dri"
Load "glx"
Load "record"
Load "xtrap"
Load "speedo"
Load "type1"
EndSection

Section "InputDevice"
Identifier "Keyboard0"
Driver "keyboard"
EndSection

Section "InputDevice"
Identifier "Mouse0"
Driver "mouse"
Option "Protocol" "auto"
Option "Device" "/dev/mouse"
EndSection

Section "Monitor"
Identifier "Monitor0"
VendorName "Monitor Vendor"
ModelName "Monitor Model"
HorizSync 31.5 - 50.0
VertRefresh 50.0 - 75.0
EndSection

Section "Device"
#Option "SWcursor" # [bool]
#Option "HWcursor" # [bool]
#Option "NoCompression" # [bool]
#Option "NoAccel" # [bool]
#Option "TV" # [str]
#Option "TV_Output" # [str]
```

```
#Option "TVOverscan" # [str]
#Option "ShadowFB" # [bool]
#Option "Rotate" # [str]
#Option "FlatPanel" # [bool]
#Option "ColorKey" # i
#Option "OSMImageBuffers" # i
Identifier "Card0"
Driver "nsc"
Option "NoAccel" "True"
VendorName "Cyrix Corporation"
BoardName "5530 Video [Kahlua]"
BusID "PCI:0:18:4"
EndSection

Section "Screen"
Identifier "Screen0"
Device "Card0"
Monitor "Monitor0"
DefaultDepth 16
SubSection "Display"
Depth 16
Modes "1024x768" "800x600" "640x480" "400x300" "320x240" "320x200"
EndSubSection
EndSection
```

There are many good reasons to switch to using XFree86 4 if you can, including:

- Xv and DGA accelerated graphics support.

- An end to the need to specify weird virtual screen sizes (the Option Display-Compression setting achieves this).

- More flexible server options.

- The default video modes have "friendlier" syncrates than the 3.x server.

- Translucent mouse cursors.

Now, that's only half the picture. Due to a bug, the XFree86.org official release code does not support scandoubled modes (for example, 320 × 240) on Geode. I have generated a patch for this bug. If you're interested in the gory details, you can find my original posting on the topic, with an explanation of the problem, at *http://www.mail-archive.com/devel@xfree86.org/msg00455.html*. Here's the patch:

```
--- Begin patch for disp_gu1.c

130a131,135
> /*
>  * Bugfix to gfx_is_mode_supported to fix problems with doublescan
modes
>  * Lewin A.R.W. Edwards <[EMAIL PROTECTED]>
>  */
>
839d843
<
840a845,850
>       int tmp_yres;
>
>       tmp_yres = yres;
>     if (DisplayParams[mode].flags & GFX_MODE_LINE_DOUBLE)
>         tmp_yres = tmp_yres / 2;
>
842,843c852,853
<       (DisplayParams[mode].vactive == (unsigned short)yres) &&
<       (DisplayParams[mode].flags & hz_flag) &&
---
>       (DisplayParams[mode].vactive == (unsigned short)tmp_yres) &&
>       (DisplayParams[mode].flags & hz_flag)   &&
850a861
>
878a890
>


--- Begin patch for nsc_gx1_driver.c

150a151,155
> /*
>  * Minor patches to allow support of low-res video modes
>  * Lewin A.R.W. Edwards <[EMAIL PROTECTED]>
>  */
>
475c480
<       { NULL, 25175, 135000, 0, FALSE, TRUE, 1, 1, 0 };
---
>       { NULL, 10000, 135000, 0, FALSE, TRUE, 1, 1, 0 };
937c942
<   minHeight = 480;
```

```
---
>     minHeight = 200;
1850c1855,1856
<     if (MemIndex == -1)                              /* no match */
---
>
>     if (MemIndex == -1)                              /* no match */
2363a2370
>


--- Begin patch for nsc_gx2_driver.c

145a146,150
> /*
>  * Minor patches to allow support of low-res video modes
>  * Lewin A.R.W. Edwards <[EMAIL PROTECTED]>
>  */
>
474c479
<        { NULL, 25175, 229500, 0, FALSE, TRUE, 1, 1, 0 };
---
>        { NULL, 10000, 229500, 0, FALSE, TRUE, 1, 1, 0 };
911c916
<     minHeight = 480;
---
>     minHeight = 200;
```

Using my patched driver enables 400 × 300, 320 × 240 and 320 × 200 graphics modes, which are useful if you need to play VideoCD or other low-resolution movie content on a Geode platform. However, you will still have to contend with the following issues:

- The NSC driver does not, apparently, fully support autoprobing. This means that running XFree86 -configure will not generate a completely valid XFree86Config file (it will "kinda" work, but it won't give you a full range of resolutions and will require some manual tweaking).

- It appears that the Geode, or at least the X driver for it, doesn't support DDC so the monitor syncrates in an auto-generated XFree86Config will be arbitrary.

■ If you're running with display compression enabled, you may see minor video glitches onscreen, particularly if your application writes directly to display memory. This phenomenon appears to be a momentary loss of sync, like a skipped v-sync pulse, and it is yet another of the problems caused by the ridiculous "video compression" feature of the CS5530. The line:

```
Option "NoCompression" "True"
```

in the Device stanza in your XFree86Config file fixes this.

■ UI rotation is supported using the Option "Rotate" "CW" or Option "Rotate" "CCW" switches. However, these will fail catastrophically unless you also use Option "ShadowFB" "true". This has a fairly severe performance downside and I don't recommend it.

■ Flat-panel support appears to be partly broken, at least on the PCM-5820 with current BIOS versions. If you need to use a direct-connect parallel or LVDS LCD, then for the time being you are probably best off using the VESA driver. Neither the vanilla XFree86 driver nor my patched driver will work correctly on most of the LCDs I have tested. The National Semiconductor server does work, but it doesn't support scandoubled video modes. (Note, by the way, that you need to specify `Option FlatPanel True` if you are using XFree86 4.x with an LCD system).

■ The nsc_drv.o driver does not correctly save/restore the entire video subsystem state with some BIOS versions. This makes it impossible to switch from X to a different virtual console. It also means that the system will lose sync and go into an undisplayable video mode if you exit X. There is no workaround for this issue at this time; use XFree86 3.x if this is a problem for you. This problem is known to exist on the PCM-5820 (all 1.x BIOS versions), Wafer-582x (all versions) and the e-valuetech EBC-3410. It does not affect the EBC-5410 with the BIOS versions I have tested to date.

4.7.5 Hybrid and Unusual Interfaces

Choosing a graphics interface method in Linux is quite complicated, because many of your possible options overlap, and certain combinations of them can coexist happily on the one system. For example, it's possible for either svgalib or X to run on top of the framebuffer device; in fact, embedded ARM-Linux systems with LCDs (such as PDAs) are almost universally implemented with an X server running on top of the appropriate framebuffer driver. Even though X is using the framebuffer, there's nothing to stop your application from writing into video memory directly and only using X for features that absolutely require it.

I'd like to share with you, briefly, two apparently little-known methods of implementing a GUI, neither of which are talked about very frequently (if at all). I've had success with both of these in commercial products, and I feel that they saved me considerable time in the applications I was implementing. Each one solves a very different set of problems.

My first suggestion is to include an embedded web browser and a simple web server on your appliance, and implement as much as possible of the user interface as web forms processed (through the web server) by a backend program. The great thing about this method is that you automatically get "free" remote control of the appliance over a TCP/IP network connection, if available. This technique doesn't work well for all types of appliances (for instance, I wouldn't try it with something like a digital video recorder), but it does work exceedingly well for implementing the configuration front-end on an appliance that spends most of its time doing noninteractive things. An example of this would be an electronic advertising sign sort of application; most of the time, it's running movies and playing still images, but occasionally the user needs to twiddle the configuration. Another good example is a machine on a factory floor, controlling some largely automated process such as counting or sorting; you might want to have a local console so that operators can perform occasional maintenance functions directly at the machine, but mostly you will want to operate it remotely.

One suitable backend for this type of system is the industry-standard Apache web server included with most Linux distributions (including Fedora). Although it's rather overkill for the type of application we're discussing, it's easy to use, it is easy to write compatible CGI modules in many different languages, and the server is pre-integrated with the OS distribution, which make it an obvious starting point, if nothing else.

Choosing a web browser to run locally is a bit more challenging. Mozilla/ Netscape is a grotesque leviathan; it's slow to start and has an enormous RAM and disk footprint. It is extremely sluggish on Geode, mostly due to the slow performance of X in general. Opera is a possibility, but it's a commercial product and it still doesn't have wonderful performance. For the application we have in mind, I recommend using either Dillo *http://www.dillo.org/*, or eLinks *http://elinks.or.cz/*. Dillo is a small, reasonably fast X-based browser, and it's particularly good at rendering pages in a cosmetically similar fashion to the "big" browsers. This may be important in applications where the browser will be called upon to render external content in addition to the local configuration pages. However, if that feature isn't overridingly important to you, I suggest eLinks as a better choice. eLinks can run either on the standard framebuffer console, or as an X application. In either case, it uses its own font-rendering engine, which leads to cosmetically different presentation than you would see with a more conventional browser. However, it does support a large number of useful features—secure connections, support for forms, some scripting functionality, and so on. Since it doesn't run with an X event model, it also lends itself admirably to being adapted for use in embedded environments that don't have traditional input devices. For example, in one application, I have adapted eLinks to use five pushbuttons for page navigation; two buttons scroll the page up and down, two buttons select "previous link" or "next link" (amongst the hyperlinks on the currently visible page), and the remaining button enters the currently activated link. Holding down that fifth button brings up a context menu that allows you to move forwards or backwards in the page history.

My second suggestion contains rather a lot of cheating. For a rather large project, I needed to implement a system that was able to run a few X applications and could also run an XFree86-based movie player application that needed to be able to change video modes and use hardware MPEG playback acceleration. However, the device needed to present a slightly souped-up version of a proprietary GUI that was originally developed on a much older product. (The older product was OS-less; it ran on a fairly low-performance 32-bit architecture with very little operating system support).

It just so happened that the proprietary GUI portion of this code was already entirely extant in the older project, so I didn't want to move it on top of an existing graphics library. I eventually developed a hybrid sort of system. The machine boots into XFree86, and launches my application. To cut down on system resource usage, there is no window manager; the startup scripts simply spawn the X server, pause for it to finish starting up, then launch my executable. My program then obtains the starting address of video frame memory, and mmap()s it into its address space (in a similar manner to the framebuffer example code I described earlier). By making a few assumptions about video memory layout, it can run the exact same code used in the older OS-less product. When it needs to provide a function that requires interaction with X, it simply spawns a subprocess; a movie player, web browser, and other similar "high-level" applications are all provided.

This ramshackle-sounding system actually works very well, and it allows the proprietary portions of the GUI to remain portable back to the older, non-PC-based versions of the appliance. Furthermore, the main application doesn't have to deal with Geode's sluggish X performance.

Quite possibly, neither of these suggestions exactly matches your system needs. The point I'm trying to make here is that there is room for lateral thinking when choosing your interface. It's entirely possible to tailor your user interface technology to the specific needs of the application you're trying to implement.

4.8 Infra-Red Remote Control in Linux Using LIRC

There are several sorts of applications where it may be useful to offer infra-red remote control capabilities. The obvious example is a homebrew DVR (digital video recorder) or TV-top player box for video content downloaded off the Internet. In an industrial or laboratory setting, however, there are other possible uses for IR control. For example, you may want to have your electronics in a sealed box to protect against environmental hazards (water, corrosive chemicals, etc). Your appliance may be mounted somewhere difficult to reach. Or you may simply want to prevent random passersby from tampering with equipment settings—authorized personnel with the appropriate remote control can still access these settings easily.

Most super-I/O chips, including the Winbond W83977 on our Advantech board, include an IR decoding function, configurable either for bidirectional IrDA communications or for receiving commands from CIR (consumer infra-red) remote controls. Note that these two functions are very different, and are supported by completely different software. IrDA is a very complex bidirectional protocol; although some remote controls use IrDA, almost any consumer remote control you're likely to acquire will use a simpler consumer protocol, such as the Philips RC5 code set. In this section, I'll show you how to set up your SBC to receive the signals from almost any arbitrary remote control. The specific remote I'm using in my worked example here is a generic cable box controller supplied by Infrared Remote Solutions Inc. *http://www.infraredremote. com/*, part IRSI-07-15-01. A sample is shown in Figure 4-2.

Figure 4-2:
Example IR remote

I chose this device to work with because it has the fewest buttons of any remote I own, thus making for a nice simple example. For convenience, I will refer to the buttons (from upper left to lower right) as 1, 3, W, A, S, D, and X—because this layout can nicely be emulated on a QWERTY keyboard with a roughly similar button layout. You may prefer to use a universal remote, in which case you can simply pick an appliance type and model (say, Sony® DVD player) and follow the remote's instruction manual to set the universal remote to emulate the controller for that device. This technique has the advantage of ongoing reproducibility; you can be fairly sure of being able to acquire a steady supply of an off-the-shelf universal remote—and even if your specific model is discontinued, you will be able to switch to another model as long as it handles the same appliance types.

Let's begin with a thumbnail description of how an infra-red remote operates: The remote control itself consists of a key matrix, an application-specific microcontroller, and an IR LED. When a key is depressed, the microcontroller generates a sequence of bursts of carrier signal, typically somewhere between 30~40 kHz—in our case, 38 kHz. The burst sequence encodes a button ID; these codes are arbitrarily mapped to appliance functions. There are several common encoding protocols, and most of these protocols include some kind of subprotocol to differentiate between multiple devices of the same type.

On the receiver side, it is normal to use an integrated IR receiver module as the front-end, rather than assembling something out of discrete components. An example of the sort of component you would find here is the Vishay TSOP12xx or Sharp GP1U series of parts. In general, these receiver modules consist firstly of an optical IR filter and photodetector. In the case of the Sharp and Vishay devices, the housing is simply molded out of an IR-transparent resin. Some older modules were constructed of metal with a small IR-transparent window at one end. This hardware is followed by, at minimum, a bandpass filter centered on the nominal carrier frequency, and a demodulator circuit that turns the carrier frequency into a solid logic level, normally HIGH (or high-impedance) for no carrier, and LOW for carrier detected. Most available detector modules have a little extra intelligence in them to reduce noise sensitivity by ignoring extremely short carrier bursts.

Note that there are two compatibility parameters here: the sensitive band of the IR detector (and the transparent range of its associated filter) must include the output wavelength of the LED in your selected remote, and the detector module's filter frequency must match your remote's carrier frequency. In practice, you will find that the sensitive band of the detector is wide enough to cover any IR LED you can purchase, so you generally don't need to worry about it. There is also a fairly wide range of acceptability in the carrier frequency parameter, and again you'll find that almost any receiver module will appear to work with most remote controls. However, the sensitive range and view angle of the sensor will be reduced, perhaps severely, the more deviation there is between your remote's carrier frequency and the receiver module's nominal frequency.

By the way, there is a significant exception to the statements I just made: In an effort to reduce error rates for high-speed data transfers, some IrDA transceiver modules filter out consumer remote signals very effectively. You may run into this issue if you're attempting to embed your application on a laptop or other appliance that already has its IR receiver built in. The only way you can work around this sort of problem is by using a different receiver module.

Tip: Sometimes while debugging you'll find yourself wondering if the IR transmitter is actually sending anything. There are IR-sensitive cards sold for detecting these kinds of emissions, but if you don't have one, you can also use almost any digital camera or camcorder with an electronic viewfinder (as opposed to a simple optical viewfinder). Just point the camera at the remote and look at the viewfinder screen; IR output will show up as a bright blue-white light. The CCDs used in consumer cameras are quite sensitive to long wavelengths; although cameras have filters in them to remove ambient IR light, the output of the remote's LED is strong enough to pierce through this filter.

Before we go any further, we need to connect an IR receiver module to our SBC. For the remote I specified, we can use the Sharp GP1UV701QS or Vishay TSOP1238 receiver, or an equivalent part (the vital criteria being +5 V supply compatibility, and 38 kHz carrier frequency). On the PCM-5820, the IR interface is CN7, which is a five-pin, 2mm-pitch single-in-line connector manufactured by JST. The correct mating connector is JST's PHR-5[28]. Hirose (HRS) makes a visually very similar but tragically incompatible connector; beware! The pinout is as follows:

Pin	Name	Description
1	Vcc	+5 V supply for transceiver or receiver module
2	NC	No connection (but see the following note)
3	IR_RX	Demodulated data input to SBC (connect to output pin of receiver if using a 3-pin receiver module)
4	GND	Ground
5	IR_TX	IR LED control signal for IrDA transmission

NOTE: This pinout is *almost* standardized. However, some other boards (for example, the BCM EBC-5410) use pin 2 for a dedicated CIR input. Advantech has chosen not to implement the dedicated CIR functionality provided by the Winbond Super I/O. If you are using a consumer infra-red receiver module on a non-Advantech board and you have reception problems, try connecting your IR receiver's output to pin 2 instead of pin 3.

Now we need to look at the software support required to make use of this detector. Before doing anything further, however, you need to ensure that the Winbond super-I/O chip on the SBC is configured for IR reception. Go into CMOS setup and navigate to the "INTEGRATED PERIPHERALS" page. Configure the following settings[29]:

- Onboard Serial Port 2: Disabled
- Onboard IR Controller: Enabled
- IR Address Select: 2F8H
- IR Mode: IrDA
- IR Transmission delay: Enabled
- IR IRQ Select: 3

---

[28] Tip: The crimp tool for these connectors is quite expensive, though the parts themselves are dirt cheap. You can either improvise with a pair of pliers or a different crimp tool (your results won't be very strong; reinforce with hot-melt glue) or alternatively salvage one from something else. In many CD-ROM drives and portable audio CD players, the connector you need is used to connect the hub motor to the main PCB. If you have a dead one of these appliances lying around, look inside it!

[29] These settings are correct for BIOS version 2.00—older BIOSes have slightly different options and a spelling mistake or two. The important features are: IrDA mode, I/O address 2F8 (COM2), IRQ 3, and ensure that the real COM2 port is disabled.

With this accomplished, we're ready to compile and install the Linux IR remote-control software, LIRC. You'll find the sourcecode archive on the CD as /linux/lirc-0.6.6.tar.gz.

LIRC consists of several components and addons, of which three are of principal interest to us. First is the kernel module that talks to the IR UART and pipes the mark-space burst data to the next overlying software layer. The lirc project supports several different types of IR interface; the one we'll be using is **lirc_sir**. Then we have **lircd**, a daemon that runs in the background and listens to the kernel module, translating the mark-space codes into a standardized data format via a configuration file that describes the particular remote control transmitter you're using. Lastly, we have **irrecord**, which is a test program used to analyze an unknown remote control and generate a lircd configuration file that will work with it.

We begin by configuring, compiling and installing the kernel-mode driver. First, extract the lirc source archive and run the configure script. Assuming the CD-ROM accompanying this book is mounted at /mnt/cdrom:

```
cd /usr/src
tar zxvf /mnt/cdrom/linux/lirc-0.6.6.tar.gz
cd lirc-0.6.6
./configure
```

In the top-level configuration dialog, select option 1 (Driver configuration) and press Enter. Navigate down to option 6 (IrDA hardware) and press Enter. Select option 1 (SIR IrDA) and press Enter. Navigate down to "COM2 (0x2f8, 3)," press Space to select it, and press Enter. You'll be returned to the main menu; select option 2 (Software configuration) and press Enter. Make sure that all five options here are unchecked, and select OK. You'll be back at the main menu once more; select option 3 and press Enter. You're now ready to build and install the LIRC module with make ; make install.

At this point, you should also edit /etc/modules.conf and add the line:

```
alias char-major-61 lirc_sir
```

Note, by the way, that it's also possible to specify options in modules.conf to override the compiled-in driver defaults. However, we already set the driver up correctly for our hardware during the configuration phase, so we don't need to add any overrides.

In order to use the IR capabilities of the serial port, we have to make sure the port in question isn't attached to the Linux serial driver. There are basically three ways of doing this: don't load the kernel serial port driver (leave it out of the kernel), unload the driver (which requires that you have built it as a module), or force it to relinquish the port we're using for IR. The last method is the simplest, and can be achieved (on the PCM-5820) with the command `setserial /dev/ttyS1 uart none`. Now, type `modprobe lirc_sir` to load the IR driver module[30]. If you get an error that the module couldn't be found, manually edit /lib/modules/2.4.24/modules. dep and add a dependency line that reads:

```
/lib/modules/2.4.24/misc/lirc_sir.o:
```

At this point, we have the basic driver infrastructure working, and before going any further, we need to teach LIRC about the characteristics of our remote control, using the irrecord utility. Run `irrecord -f /etc/lircd.conf` to start the training process. Note that this command line forces irrecord to run in a "dumb" raw mode. **Due to hardware or possibly firmware-induced glitches on the PCM-5820, you MUST use this raw mode to train a valid configuration.** If you are running on a non-Advantech SBC (or if you are performing this experiment on a laptop), feel free to omit the –f parameter. You'll get a more flexible and much simpler configuration file.

---

[30] It isn't normally necessary to load the port driver module manually like this. If you set up the devices in /dev and the alias line in modules.conf, then starting the lircd daemon should automatically load the appropriate port driver. I've detailed the process here manually so you can see immediately if there is a problem with the port driver, rather than getting a cryptic error out of lircd when you come to run it later. But it's good practice to load your expected driver manually anyway—that way you can provide more meaningful black-box information when something goes wrong.

When you run irrecord, you will first be prompted to read a couple of pages of information; press Enter twice to skip past this and start recording. In the first stage of this process, you'll be asked to hold down each button on the remote for at least 1 second. Dots will appear on the screen to indicate that irrecord is successfully receiving data from the remote control. This process continues until you've completed a full line of dots. (If there aren't enough buttons on your remote to meet this condition, you can press the same button multiple times. The important thing is to be sure you've given irrecord a good sample of the different codes generated by your remote).

Once you've completed a full line of dots, if you're not using raw mode, irrecord will proceed to the second stage of learning. Again, you should go through every button on the remote, holding each one down for at least a second. If everything is working correctly at this point, holding down each button should generate only one dot, even if you hold down the button for a considerable time. As before, this second stage continues until you have covered an entire screen line with dots.

At this point, we begin assigning names to the buttons. Simply enter a text label for the button to be "learned" and press Enter. Irrecord will prompt you to hold down the button in question, and will start listening for a code on the IR port. For some types of remote (not the specific one we're using, though), irrecord should say "Got it," followed by the message "Signal length is [integer]" for each learned button. These signal lengths should be similar for all the buttons on your remote; if you suddenly get a very short signal length (typically 1), this means that a glitch interrupted the learning process. You should re-teach LIRC that button—just enter the same button name again, let LIRC recognize it, and manually edit the configuration file to remove the erroneous entry after you've finished with irrecord. You'll recognize the bad entry because it will be very short in comparison with the good entries.

Depending on what sort of remote you were training, you may now be prompted to press a single button repeatedly as fast as you can, so that irrecord can check for toggle bits. Some IR protocols include a spare bit in each button ID code, which is toggled each time you press the button. The purpose of this bit is to detect when a continuously-repeated signal is temporarily interrupted by a physical obstacle. To demonstrate this feature in operation, point your TV remote at the set, press and hold the power button, and wave your hand in front of the remote's LED. Note that the TV set doesn't go off and on as you uncover the LED!

Note: It is *extremely important* that your IR environment is as quiet as possible while training LIRC to recognize a new remote control. Although the receiver module does have some hardware intelligence in it to filter out spurious signals, under normal conditions it is common for glitches to make it all the way through to the SBC. Fluorescent lights, including energy saver compact fluorescent bulbs and the inferno of nuclear fusion that beats in on us through unshaded windows (if you happen to be on the day-side of the terminator) are particularly evil sources of noise. If you are having trouble teaching LIRC, try darkening the room or covering the remote and IR receiver with a towel and working by feel under the towel.

If you care to examine the configuration file generated by this process, you will see a small header followed by a stanza of information for each trained button. For instance, the stanza describing the 1 button would look something like this:

```
name 1
    39    13507 39     1086  39    2222
    39    1086  39     1086  39    1085
    39    1085  39     1086  39    2223
    39    2221  39     2225  39    1084
    39    1086  39     2223  39    1085
    39    1085  39     1086  39    1085
    39    1086  39     1084  39    1087
    39    1085  39     1086  39    1084
    39    1085  39     2224  39    2223
    39    2222  39     2223  39    2222
    39    2224  39     2223  39    2223
    39
```

This represents a header pulse (the 39 13507 leadin), followed by a 32-bit button code, MSB first. Our remote uses the sequence 39, 1086 to represent a zero and 39, 2222 to represent 1 (note the slight variances in the data above; LIRC offers a "fuzziness" parameter allowing you to tweak just how much "wobble" is acceptable in the burst lengths). Thus, the actual code being transmitted for this button is, in binary,

0100 0001 1100 1000 0000 0000 1111 1111—or 0x41C800FF. (In fact, the 0x41C8 header is a vendor-specific code used to distinguish our OEM remote from other remotes using the same protocol; only the last 16 bits of the button code are actually useful data).

The long format you just saw illustrated is a very verbose way of describing the button code. In a configuration file that wasn't recorded in the dumb raw mode, LIRC simply defines what is recognized as "1," what is recognized as "0," the common prefix, if any (0x41C8 in our case) and some other information about how the button presses are encoded. It then describes each button simply with the hexadecimal number that's being transmitted by that button; 0x00FF in the case in the preceding paragraph. These sorts of configuration files are much easier to read and edit, but unfortunately the Advantech board can't work properly with them. It's unclear at this time whether this is a hardware issue or a BIOS bug, but it seems to be a BIOS issue since the exact same software configuration can be made to work on other Geode boards with the same hardware.

You should now test that your configuration file is valid by running the LIRC daemon, lircd, and then starting the "watcher" program irw. Once irw is running, aim the remote at the sensor and press a few buttons. You should see output something like this (one line of output for each button you pressed):

```
0000000000000001 00 1 /etc/lircd.conf
0000000000000005 00 s /etc/lircd.conf
0000000000000007 00 x /etc/lircd.conf
```

The four fields in this output are: the 64-bit code of the button being pressed, the repeat count, the name you assigned to this button during training, and the name of the remote control. You can edit the name of the remote in the configuration file; for example, you can rename it from the default "/etc/lircd.conf" to, say, "dvd-remote." If you then concatenate multiple lircd.conf files, lircd will recognize *all* the defined button codes and will inform you not only the code of the button that's being pressed, but also which remote control it's on. This is handy if, for example, you have trained LIRC to recognize remote controls for both a VCR and a DVD player, and you need to determine which "Play" button has been pressed.

Note that for a configuration file recorded in raw mode, the button code simply represents the position of the button description within the configuration file; the first button encountered is numbered "1," the next "2," and so on. If you are using a "smart" configuration file, the button code will be the actual binary data for the button; for our remote, this will be a number of the form 0x0000000041C8????, where ???? represents the button code. This detail isn't particularly important, however, as we will be working with the names we assigned the buttons, rather than their raw codes.

Now it's time to integrate IR support into our own application. Because of the unusual limitation of the Advantech IR hardware, I'm going to illustrate only non-repeated keystrokes. The raw configuration file we generated earlier won't recognize held-down buttons; it will recognize the first press, but not subsequent repeat events. This is roughly equivalent to having a rigorously debounced pushbutton mounted on your appliance.

In order to listen to the lircd daemon, we must open a connection to /dev/lircd. From this we obtain a regular file stream from which we can read incoming button data, in exactly the same text format displayed by irw during the tests we just performed. Let's suppose we only want to read the buttons defined for the remote control I described earlier. Following is a complete suite of functions for reading the IR stream and implementing a buffer of incoming button presses. The main limitation of this set of functions is that it only works when you have defined unique single-character names for each button on the remote.

To use these functions, first call Init_LIRC(). This function opens a connection to lircd and then clones off a separate process that continuously runs the Do_LIRC() function. Do_LIRC() doesn't chew overly much CPU time, because it spends most of its time blocked on the read operation, waiting for data from lircd. To get a good idea of how much jitter this task introduces to your system, create a main loop that strobes a bit on the parallel port, then sleeps for, say, 100 ms and repeats the process indefinitely. Put your scope on the pin of interest, with a slow sweep rate, and observe how the system behaves when you're pressing IR buttons. (Try not to have anything else running. Pagefile access, in particular, will mess up your results here). If everything is working properly, you should see that incoming IR doesn't interfere very much, if at all, with system timings.

```
// This buffer stores incoming "keystrokes".
char input_buffer[16];

// Socket for communication with the LIRC daemon
int lirc_fd;

// Stack area for the LIRC subprocess
unsigned char LIRCstack[8192];

/*
   Insert character to head of keyboard buffer and push others down
*/
void CON_Buf_Insert(char c)
{
   int i;

   for (i=1;i<sizeof(input_buffer);i++)
      input_buffer[i] = input_buffer[i-1];
   input_buffer[0] = c;
}


/*
   Check for presence of a character in buffer; don't remove it
   Returns nonzero if there is a character waiting in the buffer
*/
int CON_Count_Buffer(void)
{
   int i = -1;

   if (!input_buffer[0])
      return 0;

      for (i=0;i<sizeof(input_buffer) && input_buffer[i]; i++)

   return i;
}


/*
   Flush input buffer
*/
void CON_Buf_Flush(void)
{
   memset((char *) input_buffer, 0, sizeof(input_buffer));
}
```

```
/*
   Get key
   Returns character from head of keyboard buffer [and moves buffer
up one step!]
   or 0 if no character is available.
*/
char CON_Buf_Get(void)
{
   int i;
   char c;

   if(!input_buffer[0])
      return 0;

   c = input_buffer[0];

   for (i=0; i<sizeof(input_buffer) - 1; i++)
      input_buffer[i] = input_buffer[i+1];
   input_buffer[sizeof(input_buffer) - 1] = 0;
   return c;
}

/*
   Get key (wait if none available)
*/
char CON_Buf_GetWait(void)
{
   char c = 0;
   while(!c) {
      c = CON_Buf_Get();
   }
   return c;
}

/*
   Subprocess to communicate with LIRC
*/
int Do_LIRC(void *p)
{
   char buf[128];

   while (1) {
      int cr;
```

```
        memset(buf,0,sizeof(buf));

        cr=read(lirc_fd, buf, 80);
        if (cr > 0) {
            char key[16];
            int count;
            char *p = buf;

            key[0]=0;
            // skip serial#
            while (*p && *p!=' ')
                p++;
            if (*p) p++;

            // skip count
            count = atoi(p);
            while (*p && *p!=' ')
                p++;
            if (*p) p++;

            if (count) {
                // We ignore repeat codes. You can process them
                // if you wish.
            }

            if (*p && (count == 0)) {
                CON_Buf_Insert(*p);
            } // if (*p && (count == 0))
        } // if (cr > 0)
    } // while(1)
}


/*
    Call this function once at program startup. It connects to the
LIRC daemon
    and starts a subprocess that scans the incoming data stream.
*/
void Init_LIRC(void)
{
    addr.sun_family = AF_UNIX;
    strcpy(addr.sun_path, "/dev/lircd");
    lirc_fd=socket(AF_UNIX, SOCK_STREAM, 0);
    connect(lirc_fd,(struct sockaddr *) &addr, sizeof(addr));
```

```
    // start subprocess
    clone(Do_LIRC, LIRCStack + sizeof(LIRCStack) - 4, CLONE_VM |
CLONE_FILES, NULL);
}
```

One final note about LIRC and system performance: this module is only well-behaved if it is running on a "real" infra-red UART. If your machine lacks this hardware, and you're running with one of the homebrewed adapters discussed in the LIRC documentation, you will notice a significant drain on system resources. When using those simple bit-banged dongles, LIRC has to measure all the mark-space timings in software, which is excruciatingly CPU-intensive.

4.9 Introduction to Machine Vision Using Video4Linux

4.9.1 Acquiring Image Data from Cameras

The E-2 is equipped with several low-resolution color cameras, connected directly to the controlling SBC via USB. Besides providing interesting underwater pictures, these cameras are used for autonomous target-seeking. In this section, we'll briefly look at some of the simple machine vision concepts that I use in the E-2 project. Please note that what I'm concentrating on here is how to acquire real-time images in a Linux environment; that is, the input side of the machine vision equation. Algorithms for analyzing image data like this are explained in thick, dry and generally rather expensive books that you are welcome to acquire and study separately. Here, I'm primarily offering you the acquisition infrastructure you can use to slip the example sourcecode in those books directly into your real-time system, along with some information about pre-processing the pixel data.

For our example code, we're going to use a cheap pencam camera based on the ST STV0680 chip. The specific camera I used is a "Jazz Digi-Stix JDC11," which is available under several different names for between $10–$20. If you're looking for this exact camera, probably the easiest place to find one is eBay. There are numerous other inexpensive cameras based on the same chip. You can, however, use any video capture device that has a Video4Linux driver; the procedures are almost exactly the same.

The application side of the V4L driver API is described for you—*very tersely*—in Documentation/video4linux/API.html in your Linux source directory. Note that 2.4.x and earlier kernels implement the first version of the V4L API, referred to as V4L1. There is a new API under development—V4L2—which is going to be the standard API in kernel 2.6 and can be retrofitted to 2.4.x. At the time of writing, support for V4L2 isn't as complete as V4L1, so we won't deal with the newer API here. Unfortunately, aside from the kernel tree documentation, it's strangely difficult to find concise programming information about V4L1[31]—the definitive reference is the sourcecode for the xawtv application, which is horribly general-purpose and difficult to understand. One of my motivations for including sample V4L source here is to illustrate that video capture in Linux doesn't have to be complex (at least, not as long as you're willing to work with a constrained subset of the video capture devices supported by the kernel).

In the vidcap directory of the sample sourcecode archive, you'll find the source for a small (nongraphical) applet that tries to acquire a single frame from /dev/video0 and save it as a 24 bpp Windows-compatible BMP. During the process, various interesting information about the image capture device is displayed onscreen. BMP is a convenient file format, because it basically consists of raw RGB framebuffer data with a small header that indicates the frame size. I have created a small library containing fairly portable code[32] to read and write BMP files into flat memory arrays; you'll find this code, a demo program and some documentation in the projects/bmpdemo directory. The functions I have provided handle byte-order conversion from BMP's blue-green-red ordering to the more normal red-green-blue order, and they also automatically flip the image vertically (for odd historical reasons dating back to the halcyon days of OS/2, BMPs are stored in upside-down scanline order). My functions are, however, NOT a fully-standards-compliant BMP read/write algorithm. Do not use this BMP code as the basis of any commercial software product!

---

[31] Searching for documentation on Video4Linux will, at the time of writing, lead you almost exclusively to V4L2 reference materials.

[32] My code is "portable" in the sense that it's endianness-independent—you can run it directly on ARM, x86, MIPS, etc.—but it does assume that int is at least 32 bits.

For the remainder of this discussion, I'll assume you're looking at the sample applet sourcecode I provided (projects/vidcap/main.c) and possibly the V4L header file out of the Linux kernel. The V4L API is quite simple to use—considerably easier than other interfaces such as TWAIN, for instance. You begin by opening the desired video device (/dev/video*n*, usually /dev/video0) with open(2). Next, you use ioctl(2) to query the device capabilities and make sure it's a device you can work with. The V4L APIs we're interested in support straight stream-type video capture devices such as USB cameras, but V4L also supports radios, teletext receivers and overlay-type video capture devices that drop captured video data directly into your graphics card's video memory. We won't deal with those latter three classes of device, because the first two are irrelevant to our application, and the USB pencams I'm talking about are simpler to access and never implement the overlay type of capture method. It will, however, almost certainly be necessary for you to implement overlay capture if you're using a PCI framegrabber card, or an analog video input feature integrated into your SVGA adapter. These devices are most unlikely to support the simple read(2)-based interface.

Device capabilities are queried using the VIDIOCGCAP ioctl. You pass this ioctl a pointer to an empty video_capability structure (this structure, along with the ioctl names and other V4L constants, is defined in the videodev.h header from the Linux kernel). On return, the values in this structure are filled out to reflect the device's functionality. The two things we are interested in are the flags in type (VID_TYPE_CAPTURE must be set for us to work with the device), and the maximum capture window size in pixels, defined by the maxwidth and maxheight structure members.

Now we need to look at the image grabbing parameters using the VIDIOCG-PICT ioctl. This function fills in a video_picture structure to reflect the current driver settings. The main thing we're interested in is the palette mode – we only support the RGB24 (8:8:8) data format. If the palette parameter isn't VIDEO_PAL-ETTE_RGB24, we set palette=VIDEO_PALETTE_RGB24, pass the structure back to the VIDIOCSPICT ioctl, then re-query the device with VIDIOCGPICT again to see if the requested setting was accepted by the driver. If not, we have to abort, because we don't support conversion from the other data formats.

The penultimate step is to set the video capture window size, and query the driver state immediately after this process in order to ascertain what capture window was actually set. The reason this step is necessary is that not all combinations of palette, capture window size, and capture flags are supported by all drivers; you might not get the same window size you requested, and it's important that you know *exactly* what the driver is going to deliver because when you actually come to get image data, it arrives as an unformatted stream of bytes. So, we create a video_window struct with the x and y (starting coordinate) parameters set to (0,0) and the width and height parameters set to the maximum width and height values are returned by VIDIOCG-CAP. The remaining members of the video_window struct should be zeroed – we're not using them. Next, we pass the video_window structure to the VIDIOCSWIN ioctl, then immediately reuse the structure with a VIDIOCGWIN ioctl, which will return us the *actual* values being used by the driver.

Assuming all these gyrations were successful, we can read a frame of data out of the device simply by using the read(2) function on the /dev/video0 handle. The size of the read operation is the window width * height * 3 bytes per pixel. The format is ordered logically with the red byte first, then the green byte, then the blue byte. It's as simple as that, at least with the basic pencam drivers in Linux 2.4.x. Here's the sourcecode for our simple grabber applet:

```
/*
   main.c

   Simple V4L1 applet - Capture one frame from /dev/video0 and save
it as a BMP.
   2004-04-03 larwe created
*/


#include <sys/types.h>
#include <sys/stat.h>
#include <stdio.h>
#include <stdlib.h>
#include <fcntl.h>
#include <unistd.h>
#include <sys/ioctl.h>
```

```
#include <linux/videodev.h>

#include "bmplib.h"

/*
   Demonstration main function
*/
int main(int _argc, char *_argv[])
{
   BMINFO bi;
   BLERR status;
   unsigned char *capbuffer;
   struct video_capability vc;
   struct video_picture vp;
   struct video_window vw;
   int video, rc;

   if (_argc < 2) {
      printf("Usage : vidcap filename.bmp\n"
         "Captures one frame from /dev/video0 to filename.bmp\n");
      return -1;
   }

   printf("Opening /dev/video0... ");
   fflush(NULL);
   video = open("/dev/video0", O_RDWR);
   if (video == -1) {
      printf("Cannot open, aborting.\n");
      return -1;
   }
   printf("OK.\nGetting device capabilities... ");

   // Ascertain and display capture device properties
   rc = ioctl(video, VIDIOCGCAP, &vc);
   if (rc) {
      printf("VIDIOCGCAP failed, aborting.");
      return -1;
   }
   printf("OK.\n");
   printf("Name      : '%s'\n"
         "Channels  : %d\n"
         "Audios    : %d\n"
         "Size      : %dx%d to %dx%d\n", vc.name, vc.channels,
vc.audios, vc.minwidth, vc.minheight,
         vc.maxwidth, vc.maxheight);
```

```
printf("Capabilities : ");
if (vc.type & VID_TYPE_CAPTURE)
   printf("VID_TYPE_CAPTURE ");
if (vc.type & VID_TYPE_TUNER)
   printf("VID_TYPE_TUNER ");
if (vc.type & VID_TYPE_TELETEXT)
   printf("VID_TYPE_TELETEXT ");
if (vc.type & VID_TYPE_OVERLAY)
   printf("VID_TYPE_OVERLAY ");
if (vc.type & VID_TYPE_CHROMAKEY)
   printf("VID_TYPE_CHROMAKEY ");
if (vc.type & VID_TYPE_CLIPPING)
   printf("VID_TYPE_CLIPPING ");
if (vc.type & VID_TYPE_FRAMERAM)
   printf("VID_TYPE_FRAMERAM ");
if (vc.type & VID_TYPE_SCALES)
   printf("VID_TYPE_SCALES ");
if (vc.type & VID_TYPE_MONOCHROME)
   printf("VID_TYPE_MONOCHROME ");
if (vc.type & VID_TYPE_SUBCAPTURE)
   printf("VID_TYPE_SUBCAPTURE ");
printf("\n");

printf("Getting image properties... ");
rc = ioctl(video, VIDIOCGPICT, &vp);
if (rc) {
   printf("VIDIOCGPICT failed, aborting.");
   return -1;
}
printf("OK.\n");

if (vp.palette != VIDEO_PALETTE_RGB24) {
   // Attempt to set RGB24 palette
   printf("Attempting to set RGB24 palette... ");
   vp.palette = VIDEO_PALETTE_RGB24;
   rc = ioctl(video, VIDIOCSPICT, &vp);
   if (rc) {
      printf("VIDIOCSPICT failed, aborting.\n");
      return -1;
   }
```

```
        rc = ioctl(video, VIDIOCGPICT, &vp);
        if (rc) {
            printf("VIDIOCGPICT failed, aborting.\n");
            return -1;
        }
        if (vp.palette != VIDEO_PALETTE_RGB24) {
            printf("Device does not support RGB24 palette, aborting.\
n");
            return -1;
        }
        printf("OK.\n");
    }

    printf("Brightness : %d\n"
           "Hue        : %d\n"
           "Color      : %d\n"
           "Contrast   : %d\n"
           "Whiteness  : %d\n"
           "Depth      : %d\n",
           vp.brightness, vp.hue, vp.colour, vp.contrast, vp.whiteness,
vp.depth, vp.palette);
    printf("Palette    : ");
    switch(vp.palette) {
        case VIDEO_PALETTE_GREY:    printf("VIDEO_PALETTE_GREY");   break;
        case VIDEO_PALETTE_HI240:   printf("VIDEO_PALETTE_HI240");  break;
        case VIDEO_PALETTE_RGB565:  printf("VIDEO_PALETTE_RGB565"); break;
        case VIDEO_PALETTE_RGB555:  printf("VIDEO_PALETTE_RGB555"); break;
        case VIDEO_PALETTE_RGB24:   printf("VIDEO_PALETTE_RGB24");  break;
        case VIDEO_PALETTE_RGB32:   printf("VIDEO_PALETTE_RGB32");  break;
        case VIDEO_PALETTE_YUV422:  printf("VIDEO_PALETTE_YUV422"); break;
        case VIDEO_PALETTE_YUYV:    printf("VIDEO_PALETTE_YUYV");   break;
        case VIDEO_PALETTE_UYVY:    printf("VIDEO_PALETTE_UYVY");   break;
        case VIDEO_PALETTE_YUV420:  printf("VIDEO_PALETTE_YUV420"); break;
        case VIDEO_PALETTE_YUV411:  printf("VIDEO_PALETTE_YUV411"); break;
        case VIDEO_PALETTE_RAW:     printf("VIDEO_PALETTE_RAW (BT848)");
break;
        case VIDEO_PALETTE_YUV422P: printf("VIDEO_PALETTE_YUV422P");break;
        case VIDEO_PALETTE_YUV411P: printf("VIDEO_PALETTE_YUV411P");break;

        default:        printf("Unrecognized");            break;
    }
    printf("\n");
```

```
    if (vp.palette != VIDEO_PALETTE_RGB24 && vp.palette != VIDEO_PAL-
ETTE_RGB32) {
        printf("This program supports ONLY VIDEO_PALETTE_RGB24 and
VIDEO_PALETTE_RGB32.\n");
        return -1;
    }

    // Set window
    vw.x = 0;
    vw.y = 0;
    vw.width = vc.maxwidth;
    vw.height = vc.maxheight;
    vw.chromakey = 0;
    vw.flags = 0;
    vw.clips = NULL;
    vw.clipcount = 0;
    printf("Setting video capture window... ");
    rc = ioctl(video, VIDIOCSWIN, &vw);
    if (rc) {
        printf("Failed, aborting.\n");
        return -1;
    }
    printf("OK.\n");
    printf("Querying video capture window... ");
    rc = ioctl(video, VIDIOCGWIN, &vw);
    if (rc) {
        printf("Failed, aborting.\n");
        return -1;
    }
    printf("OK, window size is %dx%d.\n", vw.width, vw.height);

    // Allocate RAM for capture buffer. At MOST we need 32 bits per
pixel.
    capbuffer = malloc(vw.width * vw.height * 4);
    if (capbuffer == NULL) {
        printf("Error allocating memory for capture buffer!\n");
        return -1;
    }

    // Capture an image using the read() interface. This is potential-
ly unsupported.
    printf("Reading buffer... ");
    if (vp.palette == VIDEO_PALETTE_RGB24)
        read(video, capbuffer, vw.width * vw.height * 3);
```

```
    else if (vp.palette == VIDEO_PALETTE_RGB32)
        read(video, capbuffer, vw.width * vw.height * 4);
    printf("OK.\n");

    printf("Closing /dev/video0... ");
    close(video);
    printf("OK.\n");

    // For RGB32, we need to convert the image down to RGB24.
    if (vp.palette == VIDEO_PALETTE_RGB32) {
        // BUGBUG - Not implemented yet!
        printf("CONVERSION NOT IMPLEMENTED.\n");
        return -1;
    }

    // Create overlying bitmap structure
    bi.width = vw.width;
    bi.height = vw.height;
    bi.bitmapdata = capbuffer;

    // Open output file
    printf("Saving output file... ");
    fflush(NULL);
    bi.fd = open(_argv[1], O_WRONLY | O_CREAT | O_TRUNC, S_IRWXU);
    if (bi.fd <= 0) {
        printf("Can't open output file.\n");
        return -1;
    }
    status = BL_Save_Bitmap(&bi);
    close(bi.fd);
    printf("Returncode %d\n", status);
}
```

4.9.2 Detecting Object Edges

Now we've got the raw image data, what do we do with it? Most machine vision tasks involve recognizing and locating shapes. In order to recognize shapes, you first need to locate the edges of any objects in the image. The simplest approach to this problem is to regard each scanline of the image as representing a section of a continuous function (grayscale pixel level), and take an arithmetical first derivative of this function to yield an "edginess plot" of the image. Various further processing can be done on this in order to generate a greatly massaged image ready to feed into your artificial intelligence system.

All the above is a bit of a mouthful, so let's look at a couple of practical examples, with sourcecode. The sample programs I'm going to talk about here don't work directly with V4L, because it's a bit inconvenient to test with known-to-be-interesting test images when you need to aim a camera at them. So, for portability and ease-of-testing reasons, the sample programs work with uncompressed 24-bit BMP files. In this text, I'll be referring to specific sample images I acquired myself; you can find all these images in the projects/images directory. You will also find the output images shown in the text. The files on the CD-ROM are the raw output from the sample code shown here; for print production reasons, what you see on the page here has been post-processed some more to improve contrast.

Let's first run a simple left-to-right scanline derivative on the image, to identify areas of change (edges, in other words). A demo program to achieve this is located in projects/bmpdemo-derive. The meat of it is the following function:

```
void DER_DeriveScanline(unsigned char *pixels, int width)
{
    int i;
    unsigned char p1,p2,result;

    for (i=0;i<width - 1;i++) {
        p1 = *(pixels+1);
        p2 = *(pixels+4);
        if (p2 > p1)
            result = p2 - p1;
        else
            result = p1 - p2;
        *(pixels++) = result;
        *(pixels++) = result;
        *(pixels++) = result;
    }
    // Put dummy column on RH side
    *(pixels++) = 0;
    *(pixels++) = 0;
    *(pixels++) = 0;
}
```

This snippet of code works from left to right on a single scanline of the input image, and replaces each pixel P with the absolute value of the arithmetic derivative of the original image contents at P and P+1. We take the absolute value for two reasons; firstly, we don't care if we're detecting a transition from light to dark or dark to light (an edge is an edge to us), and secondly, storing the absolute value obviates the need for a wider data output storage type (with a sign bit). The algorithm ignores the red and blue components of each pixel; the demo program, by the way, performs a color-to-grayscale averaging before running the above algorithm. Note that there is one unavoidably incalculable pixel at the right-hand edge of the scanline. This dummy pixel is set to 0 for convenience in later analysis.

Here's a picture of my apartment wall and attic door, before and after processing with the command `derive wall-flash.bmp wall-flash-1.bmp 1` (more about the "1" parameter in a moment, just for now suffice it to say that this additional parameter ensures that the output you see is the result of the above arithmetic differentation only). I chose this subject material because it's a relatively clean, noise-free environment illustrating a number of interesting points.

 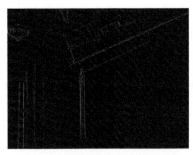

The image above was taken with bright (xenon flash) light. Let's look at an almost identical picture taken with ambient lighting only, and subject it to the exact same processing:

You'll note that although the color balance and overall brightness of the two "before" images are very dissimilar—and there's a very noticeable pattern of concentric brightness rings centered around the ceiling light in the second picture—he differentiated images look almost the same. This very simple algorithm has therefore helped us pick out edges in the image, and simultaneously erased some of the absolute effects of variable lighting. It's hard to establish quantifiably just how much the latter problem has been mitigated, because there are other factors at work in the camera. For example, in low-light conditions the exposure is automatically increased, which means that bright spots tend to get over-saturated (losing all detail) and mechanical vibrations have more chance to blur the acquired image.

To help you better visualize what's going on, I've also included some simple edge amplification code in the derive program, and that's what the third command-line parameter controls. The way this amplification works is as follows: The entire output image is scanned to determine the brightness b of the brightest pixel it contains (i.e., the sharpness of the sharpest edge in the original image). This value is then subtracted from the maximum possible pixel value (255) to obtain a "brightening factor" $f = 255 - b$. Each pixel of the image is then analyzed with a user-supplied amplification constant a. (This is the third parameter on the command line). The analysis rules are:

- If the pixel is dimmer than b/a, then it is left unaltered.

- Otherwise, the brightening factor f is added to the pixel.

The net effect of all this is to make the very brightest edges stand out from the rest of the crowd. At the same time, the relative brightness of all pixels is preserved; if a given pixel was brighter than its neighbor before processing, it will still be brighter after processing.

Now, one thing you will observe about all the "after" images is that the horizontal scan process can only detect edges with a vertical component. The horizontal crossbars of the moldings on the wall have therefore disappeared. I won't illustrate it here —you can check it for yourself easily enough using the program in the bitmap-derive2 directory, but obviously a vertical-scanning derivative is equally as blind to vertical lines as the horizontal scan was to horizontal image features. One possible solution to this problem is to take *two* passes over the initial data, one scanning vertically and one scanning horizontally, and then average the two resultant bitmaps. This approach takes somewhere between two and three times the processing horsepower of a simple one-dimensional derivative, but it will capture a lot of detail you would otherwise miss.

If you run the svgacap program described in Section 4.7.3 (which uses the preceding code) and tinker with the noise threshold parameter, you will also quickly observe that simply scanning for sharp brightness transitions isn't the best way of identifying edges. A more refined approach is to look for sign changes in the derivative. I haven't included example code to do this, but you can easily modify the simple left-to-right scanning algorithm in projects/bmpdemo-derive to give you an idea of the data that can be pulled out of the noise using this approach. This second derivative algorithm is the basis of most simple shape-recognition algorithms, because it is a true method of detecting edges.

It is interesting to run the sample programs described here over a variety of source material and observe the output, because it gives you an instructive window into the ease with which various classes of image data can be algorithmically analyzed. I particularly encourage you to tinker with various processing options using some of the navigation and hazard avoidance camera images taken by NASA's MER-A and MER-B Mars rovers; you can get some feel of just what the onboard software in those robots is seeing. The homepage of the Mars rover project, where you can download high-resolution raw image data from both robots, is *http://marsrovers.nasa.gov/home/index.html*.

4.10 Customizing Your BIOS—The Structure of a Modern BIOS

All of the SBC vendors you'll deal with will offer you the opportunity to purchase a custom BIOS. Charges for this customization range from free to more than ten thousand dollars up-front, plus (generally) a per-unit premium to preload your special BIOS version before shipping boards to you. Here are four of the most common reasons why you might want a customized BIOS for your product:

1. **Custom CMOS default settings.** Most applications that use an embedded PC will want or need to have nondefault settings; for example, no halt on keyboard error at boot time, a boot device order that doesn't start with the

first floppy drive, and so on. It's a lot safer to have these settings locked into flash than simply to set the board up once at production time and rely on the CMOS battery to retain those parameters. If your system ships without a keyboard or display that would allow the user to reconfigure lost CMOS settings, then it's **absolutely mandatory** that you burn customized defaults into flash; anything from a brief power glitch to random software errors could corrupt the CMOS contents and leave the system unstartable.

2. **Security.** You might want to lock out access to the system's BIOS setup utility by setting a nonerasable password. This is very common in set-top-box Internet appliances based around generic PC hardware.

3. **Cosmetic issues.** Hiding the normal POST messages can conceal, to some degree, the embedded-PC nature of the product. Many vendors also choose to insert a custom boot logo or message, often including a company URL for product support.

4. **Functional issues.** Examples of this might include special BIOS extensions for network booting, a ROMless SCSI card, or some proprietary piece of hardware. Perhaps your BIOS needs to be updated with custom video parameters for an LCD you're using. You might also need to tweak your system BIOS to work around some specific compatibility problems, as described below.

I'll base this section around an anecdote about the PCM-5820 that illustrates nicely why you might be forced to poke about in the internals of your system BIOS, no matter how reluctant you are to do so. The following discussion specifically describes the Award Modular BIOS, but the basic ideas are applicable to most modern BIOS vendors; the tools and exact usage procedures differ slightly, but the available options are very similar. Although this particular issue may never be a problem for you, it's an excellent way to discuss the options and procedures available for in-house BIOS customization.

When we started to use the PCM-5820, we tested and qualified our application on the then-current BIOS version. We continued to ship products based around the board for about two years, and over time Advantech revised the BIOS on several occasions. We noted the new versions, and did some basic compatibility checking, but nothing untoward happened, and we continued to ship goods without interruption.

Now, you should note that there is generally no proactive notification of upcoming BIOS revisions. It's possible to be notified of upcoming *hardware* revisions if you're savvy enough to request such notification, but you only find out about new firmware revisions when the next shipment of boards arrives. This annoying behavior, by the way, is by no means unique to Advantech—it's fairly consistent across all the vendors I've worked with. The only way you can assure a continuing supply of absolutely identical boards is by getting the vendor to set up a custom part number for you. Not only does this usually result in a slightly higher unit cost, but there are obviously large volume requirements as well.

Our application's volumes didn't justify a frozen, guaranteed-identical SBC supply chain, so one day we received a shipment of boards loaded with BIOS version 1.23, and suddenly the world fell apart. Using this BIOS version under certain circumstances[33] has the interesting property that, on approximately three attempts out of twenty, XFree86 does not initialize the video output correctly; although the computer is still running happily, the VGA port is outputting bizarre syncrates to which no known display can sync. A reboot is required to fix this condition.

Our first, obvious course was to try backleveling these new boards to an earlier BIOS version. Unfortunately, we were thwarted in this, because the release of the V1.23 BIOS update coincided with a hardware spin, too. Due to supply or price issues, Advantech had switched from using an Analog Devices audio codec to a nominally pin-and-register-compatible Realtek chip. "Nominally compatible" is of course vendor patois for "subtly different in irritating and show-stopping ways," and the specific irritation in this case is that BIOS code intended for the Analog Devices chip will play back audio noticeably too fast on the Realtek chip.

---

[33] If this problem hits you, one workaround is simply to install 128 MB of RAM. It's not completely clear why this affects the visibility of the problem, but we have performed a *lot* of testing with various configurations, and determined, with help from Advantech, that the issue doesn't appear with any BIOS version when there is 128 MB or more of RAM installed. It's pretty clearly a race condition of some kind, probably to do with a SMM interrupt occurring during the CRTC initialization code in XFree86, and adding RAM just changes latencies enough to "fix" the problem. This is an evil workaround and I promise to wear a mask of shame for even mentioning it, but a proper fix is very difficult to engineer without good support from the original BIOS supplier, and it isn't even certain who that really is.

The net result of this is that we were in a catch-22—BIOS V1.23 or later has the video initialization bug, and any BIOS earlier than V1.23 on a board that originally shipped with V1.23 or later results in incorrect audio playback speeds. After some considerable testing, we found that our XFree86 problem was shown by BIOS versions 1.22 and onwards, but *not* by V1.21, and so we needed to determine what exactly had changed between V1.21 and V1.22. Officially, the only change was a "minor change to reduce serial port interrupt latency."

There's quite a lot of exploring we can do in the board's firmware without actually having to fire up a debugger and start poking around in the code. Modern mainboard BIOSes are no longer just a single monolithic block of executable code with an entry point at FFFF:0000h; they're a distinct mini-filesystem and are practically complex enough to be considered mini-operating systems in their own right. We'll look at the Award Modular BIOS (used on the Advantech Geode-based boards— and in fact, all of the SBCs I mentioned in the compatibility table in Section 2.5), but the general ideas are the same for all modern BIOSes. The BIOS image contains a small bootstrap program, a relatively large filesystem area full of compressed code and data modules, and a small amount of meta-information describing those modules. At power-on, the bootstrap code decompresses the main system BIOS into shadow RAM, and then decompresses the various other modules into other areas (usually in the high-memory area between 640K and 1 MB)[34]. Typically, a highly integrated board like the SBCs we're using will include VGA BIOS, a network boot ROM, and perhaps some other extensions (for example, a SCSI BIOS extension).

The official utilities required to work with Award BIOSes are three DOS applications called AWFLASH, CBROM, and MODBIN. AWFLASH is the flash-upgrader utility, made available to the general public so they can update their motherboards. You can download this from Advantech's site, among others. CBROM is a utility that can decompose a BIOS image into its component modules or gather up a list of specified modules into a complete BIOS image. MODBIN works on the main system

---

[34] This is just a general outline of what happens. The detailed mechanics of a specific situation may vary somewhat. For example, some of the ROM contents may already be decompressed, and the bootloader may simply point some chipset register or interrupt vector to the data in ROM.

BIOS code module, editing the CMOS setup menu tables and a great deal of other configurable information that the user would not normally get to modify; special CPU flags, floppy drive step rate, IDE timeouts, the BIOS version string shown at boot and so on. MODBIN and CBROM are theoretically supplied to OEMs only, for factory customization. In practice, they are readily available and apparently their distribution is not actively suppressed. The impression I have formed is that the BIOS vendors don't particularly care about keeping these programs secret, but they don't want to deal with the technical support effort of releasing them officially.

In any case, there also exists an open-source project called AwardMod, which performs the same general functions as CBROM. The main differences are: (a) it can legally be distributed (I have included it on the CD-ROM with this book), and (b) being a Windows-based GUI program, it is somewhat easier to use than the Award command-line utilities, though it is definitely a piece of hacker software, and rather idiosyncratic. Here's a screenshot of AwardMod with the Advantech V1.21 BIOS loaded:

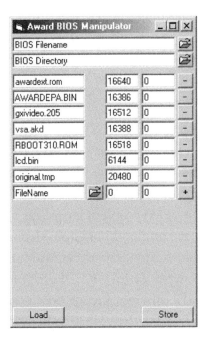

AwardMod can work with either a single-file BIOS image, or with a decomposed set of BIOS components inside a directory. To load or save an image file (suitable for flashing onto a board with AWDFLASH), enter the path and filename in the "BIOS Filename" box (or use the browse button next to it) and click Load or Store. To load or save a set of components, enter the path to the desired directory in the "BIOS Directory" box (or, again, use the browse button next to it) and click the Load or Store button.

From top to bottom in the screenshot, the modules of Advantech's V1.21 BIOS for the PCM-5280 are: the PnP extension, graphics data for the Energy Star logo shown at system boot time, the VGA BIOS, the Geode Virtual System Architecture (VSA) code, the network boot extension, a small data structure describing timing parameters for the attached LCD, and the main system BIOS code. The number in the second column is a "magic number" identifying what kind of data is in the module. Some documentation describes this field as the target segment address for the module. This may have been true once, but it appears to be obsolete information; now, it seems that this number is purely magic (i.e., arbitrarily chosen from some unpublished table). Note that the filenames of the components are of no importance, except to tell a viewer what's in the file—the bootstrap code identifies the function of each component using the magic number. You will find that the same module will frequently have different names in different BIOS versions.

By dumping out the componentized versions of V1.21 and V1.22, we could immediately eliminate the video BIOS and network boot ROM from consideration in our problem, because they were byte-for-byte identical between the working V1.21 and broken V1.22 BIOSes. By swapping V1.21's modules one by one into a V1.22 image, we determined that our video problem lay in the VSA code. Unfortunately, the hardware-specific code that handles the audio codec is *also* in this module, because it contains all the audio and video virtualization routines. So, transplanting a fix for the video problem from V1.21 into V1.23 unavoidably brought in broken audio code. Eventually, we found a different Geode board that used the Realtek audio codec chip, merged its apparently bug-free VSA code into the BIOS for the PCM-5820, and we had a working system again. It's vital to perform thorough testing of *all* board features when you do a breathtaking piece of hackery like this, however—we found several VSA dumps that appeared to work, but subtle problems appeared due to small implementation differences—for instance, in one case the MAC address of the on-board Ethernet port would come up as 00:00:00:00:00:00, and in another case the USB ports didn't work correctly. Rigorous testing for these kinds of issues will prevent embarassment and/or extra costs for you later on.

What is this VSA code, anyway? It's a large block of emulation routines, code that runs in the "super-supervisor" System Management Mode (SMM) of the Geode processor. Among other things, it handles special I/O processing for VGA text modes (which aren't supported in hardware by the CS5530 video chip), SoundBlaster emulation, and other many other simulated "virtual hardware" features. I've received conflicting information on this point, but it seems that some SBC vendors, at least, buy their VSA code as a "black box"—if they encounter a problem in the field, they pass a report up the line to National, and when an updated version of the VSA code arrives, *deus ex machina*, it gets merged into the shipping BIOS—presumably after some testing. The downside to this arrangement is that if the VSA code is causing you a problem, you can't get "real" technical help, because almost nobody in the known universe understands the VSA code's timing issues and register-level programming of the Geode chipset. Perhaps we would have been able to fix our problem by switching off some chipset feature or altering a timing value before starting XFree86, but in the end the solution we found was to go through all the hoops above.

Hopefully, this anecdote has taught you two things. First, *no* firmware change is insignificant, and you should try to freeze the firmware you ship in your products, whenever feasible. Second, unless your volumes are enormous enough to justify a custom line-item from your vendor, the vendor will seem to work actively to defeat this goal, so you have to be prepared to do some potentially time-consuming legwork from time to time. The story I described here is close to a worst-case scenario (the absolute worst case, of course, being a problem that requires you actually to disassemble, reverse-engineer and patch the BIOS code or data structures). Hopefully, you will never need to go through a situation like the one we encountered. However, if you are shipping an embedded PC application, you more than likely *will* need to alter LCD parameters, default CMOS settings, etc. and it's important you're aware of the tools that are available to you, because the SBC vendors will want to charge you special fees for these services.

Note that for copyright reasons, I can't include any of the vendor-supplied BIOS customization utilities on the CD-ROM with this book; I can only point out their existence and demonstrate to you what kind of things they can do. However, these utilities are readily available by searching on the Internet. Unquestionably the definitive jumping-off point is *http://www.biosmods.com/*, which carries many versions of the customization utilities for popular BIOSes for free download. Pay careful attention to the versioning information supplied with these utilities. Although the program will usually perform fairly thorough version-checking when loading a BIOS image, there are so many subversions and sub-subversions of BIOS code, each of which is virtually a custom product, that caution is advisable.

CHAPTER 5

Encryption and Data Security Primer

5.1 Introduction

It is impossible to build a trustworthy control network unless the topic of security is addressed and designed into the product from the beginning. Whether you are designing a system for your own use, or for installation into some industrial or commercial application, you will need to consider how to protect it against some level of attack from the outside world, and how to protect recorded data from theft or forgery.

Although data security involves physical, procedural and other holistic aspects, most security techniques in consumer and commercial applications are centered around adding encryption to existing protocols and data formats. This is primarily because encryption is cheap, being provided by "free" software, and it is also much easier to force users to run a "secure" version of a program (with encryption features forced to be on) than it is to get them to change their data security habits. Note that encryption technology really embraces two related topics: protecting valuable data from being intercepted and read by people who aren't entitled to read it, and authenticating transmissions so that commands from untrusted sources can be identified and ignored. The latter task involves encoding or wrapping data from a trusted source with a layer that cannot be forged by a third party. It *doesn't* necessarily involve encrypting the actual data being transmitted. Be sure not to confuse these two points.

When considering measures to protect your data, you must take account of the following factors:

- What part of the data needs to be protected. In many applications, a considerable proportion of the data throughput doesn't need to be protected; only a small core of data needs protection. In other cases, it may be necessary to use different levels of protection for different classes of data.[35]

- What types of attack you need to protect against.

- Resources available to you. This includes any special restrictions on your system; power or duty cycle limitations, available CPU horsepower, and so on.

- Resources available to your potential attacker. This is usually a function of the monetary value of the information being protected. Exceptions to this rule exist, of course; for example, disgruntled ex-employees or malicious hackers may be willing to dedicate enormous time and in some cases stolen distributed computing runtime.

Note that encryption algorithms are politically hot discussion topics. Many jurisdictions have, and occasionally even enforce, laws that either prevent consumers from using certain encryption technologies, or restrict the strength of the algorithms that can be used. Some of these laws are intended to regulate traffic in "armaments," i.e., encryption technologies that could be used by an enemy. (The United States, which was once a fierce defender of laws in this category, has largely relaxed its requirements. It used to be illegal for a US citizen to sell or disclose most encryption technology to any noncitizen. Now, it is only illegal to provide these technologies to embargoed destinations).

The other class of encryption-related laws is intended to enforce intellectual property rights. The best-known golem among these laws is the United States' Digital Millennium Copyright Act (DMCA), although some other countries have or are proposing similar legislation. Amongst the numerous provisions of the DMCA, it is now a crime in the United States to disclose more or less any information about

---

[35] For example, if you were implementing a secure email system, you might want the entire message (including routing information) to be illegible to people listening on the wire. However you would need to make the routing information accessible to mail delivery software at each end of the connection. You wouldn't want to allow such systems the ability to decrypt the message body, though.

certain proprietary technologies that are used for copy protection[36]. Regardless of the original intentions of such legislation—I find them suspect at best—the net effect of these laws is to inhibit free discussion of such cryptosystems. For a practical example of this, you need look no further than the debacle about DeCSS, the encryption system used on commercial DVDs.

The upshot of all this is that it's potentially controversial, and hence inadvisable for me to include strong encryption sourcecode with this book—so I haven't. However, this should not be a serious impediment: you can simply use your favorite web search engine to find "xxx algorithm sourcecode" and you are guaranteed to find exactly what you want.

Now, any reference you read on encryption technologies will make the following assertion, and I'd like to reinforce it in your mind: **Security through obscurity is an illusion**. What this means is that any system that bases part of its "security" on the fact that the system's structure itself is secret, is fundamentally flawed. It should be assumed, even for relatively low-value applications, that any attacker has complete knowledge of the algorithms and procedures in use. The reason this is practically always true is very simple: If your application is high-value, high-security, there is a financial incentive for people to discover how it works, no matter how secret and proprietary it might be. On the other hand, if it's a low-value application, you're probably using a standard commercial product to protect it, and commercial products are sold in such large volume that they should be assumed vulnerable to some type of "script kiddie" attack—that is, an automated attack program written by one knowledgeable person, but widely distributed and easily operated by a novice. The encryption used in the password protection feature of many common archiving programs is a fairly good example of this.

Philosophy aside, in a good cryptosystem the only "key" to decrypting a given block of data is the secret key that was used to encrypt it, or an equivalent related secret that is only known by authorized persons. Any approach to security—and this extends beyond encryption, by the way—should start with the assumption that a po-

---

[36] This isn't exactly the letter of the law, but it's essentially how things stand. Worse still, it's effectively almost a worldwide law—if you perform perfectly legal reverse-engineering in, say, Europe, then visit the United States, you could be arrested.

tential attacker is fully informed about the system architecture. They will quite likely even have sourcecode to the software you are using. To use a physical-world analogy, relying on algorithm secrecy is like hanging your front door key from the doorbell, but concealing the lock so that a potential thief can't work out where to put that key.

On a closely related note, others (particularly vendors of proprietary encryption products) will argue with the following statement, but I stand by it nevertheless: Any closed-source product or proprietary algorithm is inherently insecure. It is at best very difficult to perform rigorous analysis on such products; generally speaking, it's impossible. The security of a given cryptosystem can only be proven mathematically up to a point; a much more effective proof is to document exactly how the system works and let the world of professional cryptanalysts beat on it, trying to break it. A system that withstands expert public scrutiny will withstand private attack. An algorithm that doesn't attract any expert scrutiny when released to the public's gaze is probably not innovative or contains obvious flaws; why use it when well-tested algorithms exist? Furthermore, even secure encryption algorithms can be rendered totally ineffective by implementations that leak information an attacker could use to deduce the encryption key(s).

Note, by the way, that when I use the word "cryptosystem," I'm referring to a much larger concept than simply the encryption algorithm. Merely selecting a robust encryption algorithm does *not* a secure system make, absent careful scrutiny of the entire system and the paths your data can take in, through and out of that system.

As an example, I was once called upon to work on a piece of commercial encryption software that comprised two principal layers[37]; at the bottom layer, the computer on which this software was installed had its entire hard drive encrypted at a sector level with a weak proprietary algorithm (to prevent simple text searches from finding directory information). At the top layer, the user had the option of superencrypting specific files with DES, which at the time was considered sufficiently secure for the type of information being protected. Unfortunately, this system was relatively easy to break, to one degree or another. Because the structure of a DOS-formatted disk contains many snippets of data with meanings defined by the operating system,

---

[37] These "layers" refer to crypto layers only. The software itself had numerous modules, interlinked to make it difficult for users to accidentally uninstall or bypass the product.

the unencrypted contents of these areas can be guessed by an attacker. Thus, it was easy to penetrate the lower level of the encryption system with a known-plaintext attack. A lot of potentially sensitive information was then immediately accessible, unencrypted, in temporary files and the Windows paging (swap) file. In early implementations of the program, searches through the paging file could even occasionally find the original encryption key, in plain text, exactly as the user had typed it into the key-request box when encrypting or decrypting a file.

An even more blatant example of insecure implementations can be found in a certain Windows-based encryption program (no longer on the market) from a well-known software publisher. The product in question implements several standard algorithms—DES, 1024-bit RSA, and a couple of others. The implementations of these algorithms are likely to be textbook-correct. However, the product is, by default, configured to store user keys in a keyring file. This file is password-protected; it is encrypted with a one-way hash of some user-selected password. The problem with this arrangement is that the security of the entire system hinges on the security of the hash algorithm and the algorithm used to encrypt the keychain. For unknown reasons[38], the software developer chose to use only a 32-bit key to encrypt this critical data file. Recovering the entire store of keys could easily be accomplished by brute force; thereby unlocking all the user's files despite the fact that they were encrypted with "secure" algorithms and fairly large key lengths.

The latter example is an obvious example of high security algorithms defeated by low-security key management. Unfortunately, not all such exposures of sensitive key information are so easy to detect. It is frequently rumored that (insert the name of your favorite encryption software here!) has been deliberately structured so that it leaks a few bits of key information here and there, in such a way that a person with special software can examine several messages sent by you and thereby recover your entire key. It's practically impossible to refute these arguments convincingly without full public disclosure of the sourcecode. So, I'm going to state a personal dogma: All closed-source encryption products should be regarded as potentially relying on

---

[38] Conspiracy theorists would speculate that the NSA or some similar body coerced the software publisher into making the product easily breakable. You'll hear a lot of conspiracy theories like this if you do any cryptographic work. Some of them are accurate.

"security through obscurity" to some degree. It is impossible to prove their implementation to be secure, and hence you should only trust encryption software for which the full sourcecode is made publicly available. The **only** exception to this rule—and it's a partial exception at best—is that if this closed-source software implements some known algorithms, you can compare its ciphertext output with the output provided by a textbook implementation of the algorithm, operating in the same mode, with the same plaintext input and key. You should perform such testing with a wide variety of random data. Don't use industry-standard test vectors, or vectors supplied by the software vendor—the software might be designed to detect these special cases and "play it straight" because it knows it's being scrutinized. By the way, I do *not* mean to imply that any crypto product with an open-source license is trustworthy— it's quite possible to imagine that a skilled cryptographer could hide a subliminal key escrow channel in his code that you simply couldn't observe by simple examination, or even detailed analysis, of the sourcecode. (Again, practically every popular encryption algorithm—particularly algorithms approved or recommended by government bodies—has had accusations of this nature leveled against it). The point is that it's much harder to hide dirty laundry of this kind in an open-source product.

If you're starting to become suspicious and paranoid at this point, then congratulations — and welcome to the world of data security. I'd offer you a drink, but you probably won't trust me enough to take it.

5.2 Classes of Algorithm

In the overall context of a complete cryptosystem, there are several types of algorithms which you may need to use in order to achieve a specific blend of features. Probably the most familiar type of cryptographic algorithm is the symmetric-key cipher. The ancient and venerable DES encryption standard is an example of this type of algorithm. Its chief characteristic is that there is a single secret key which must be known to both the author and recipient of a message. For many (but not all) symmetric-key cryptosystems, there is a single transformation function which performs both the encryption and decryption tasks. If we take a data block D, apply the transformation function F with key K, yielding an encrypted data block D', we can take D', run the same transformation (with the same key) over it, and get D back again.

Symmetric-key ciphers are usually fast, and generally are selected for high-bandwidth bulk data transfers. One major downside to these algorithms, however, is the need for both parties to know the secret key K. If you want to talk to someone securely, somehow you need to get the key to them without anyone eavesdropping on the conversation. Clearly, it's impractical to communicate the key in the clear (unencrypted) over your regular communication channel; if it was secure enough for such traffic, you wouldn't need to have this additional cryptosystem in the first place. Ultimately, you need to establish some secure channel (bonded couriers, for instance) to deliver the secret key material, and this is an expensive and difficult task.

Asymmetric-key algorithms solve this problem by splitting the key into two halves, referred to as the public and private keys. Any data encrypted with the public key can only be decrypted with the private key, and vice versa. The key generation mechanism is devised so that it is computationally unfeasible to calculate the private key from the public key. The beauty of this system is that you and your friend can give each other your public keys over an insecure channel, and not worry about eavesdroppers. When you send a message to your friend, you encrypt it with his public key. The only way it can be decrypted is with his private key, which only he knows. Similarly, his replies to you are encrypted with your public key, and only you are privy to the corresponding private key.

Other more or less special-purpose algorithms exist. For example, there is a class of shared-secret algorithms where the decryption key is broken into a number of parts. The algorithm is designed so that the complete key can be reconstituted by bringing together any m of n total parts, where m and n are selected according to the customer's needs. Such algorithms are typically used, in the commercial world at least, for escrowing keys to information that must be kept secret from everybody in the company, but which is critical to the business and must be recoverable if something happens to one or more of the few people who know it. For example, if you work at a company that requires you to encrypt all your data with a key that you keep absolutely secret, they might implement a two-of-three shared secret system; one secret (A) will be known to both you and MIS, one key (B) will be your private key, known to you alone, and one key (C) will be known to MIS only. With this system, you normally use keys B and C to encrypt your files. If you leave the company and don't tell anyone your key, MIS can still recover all your files by combining keys A and C. Your co-worker in the next cubicle won't be able to look at your files because he only knows A (and maybe not even that); he has his own private key B', which won't help him get into your data, and he doesn't have the MIS master key C.

Also essential for many cryptographic applications, although not an encryption algorithm in itself, is a secure random number generator (RNG). "Secure" in this context means that the RNG generates a stream of output bits which are entirely unpredictable. Among other things, this means that observation of even an infinite number of output bits will not give the viewer any ability to predict the next bit. Further, the distribution of bits should be perfectly uniform; good random data is white noise. Unfortunately, computers are deterministic state machines—there is no way of generating a stream of truly random bits in software alone. The best that can be done is to generate a pseudorandom sequence, which repeats after some long interval. The cornerstone of a cryptographic implementation that relies on pseudorandom numbers is finding some truly random "seed" information to select an arbitrary starting position in the pseudorandom sequence. Some programs use the user's keystroke latencies; some use real-time clocks, and so on. Ultimately, none of these methods (alone) is secure enough to be relied upon; hardware solutions must be sought if possible (for example, recent Pentium processors have a good hardware RNG built into the chip). If you can't add true random number hardware, then a reasonable second best is to combine several sources of potentially random information to obtain your seed. RSA Laboratories publishes a variety of interesting information on this and other topics; their papers are well worth reading. You can visit their web site at *http://www.rsasecurity.com/rsalabs/*.

Asymmetric-key systems, mentioned earlier, can be used to perform message authentication in addition to simple encryption. In order to achieve this while still leaving the message in plaintext (often a requirement for digital signature algorithms), it is necessary to have another class of algorithm—a secure hashing function. A good hash function will generate very unpredictable output for a given change in input bits. You can think of it as a very good pseudorandom number generator where the message to be transmitted constitutes the seed.

In the next few sections, we will apply simple analysis techniques to a few common data security scenarios, to suggest cryptosystems that are appropriate to the task.

Please note that the following suggestions are not exhaustive—there are many ways to skin a cryptographic cat. The aim is to show you the sort of thinking you'll need to do in order to pick a good match of cryptographic technology for a particular job.

5.3 Protecting One-Way Control Data Streams

Let us consider a remote-controlled hobbyist aircraft, or more specifically the link between the control box and the vehicle itself. In this application, the data to be protected is a relatively low-bandwidth stream of control information. The real-time characteristics of this are very important; if control information is delayed, the craft will probably crash. Because the aircraft has weight restrictions (and by implication power restrictions), we can also safely assume that onboard computational resources available will be limited. Similarly, the control box is likely to be handheld and battery-powered, so it will also have computational limitations. The potential attackers we can anticipate are people who want to subvert the control stream and either steal the aircraft or simply make it crash. Our likely attacker will, at best, have a laptop computer or other relatively low-power computing appliance to attempt his attack (although it's not inconceivable that someone could have a wireless Internet connection and use a distributed computing attack, it does seem very unlikely that anyone would go to this trouble).

A few other pertinent facts about this system are as follows:

- Before launching the aircraft, we can establish a known secure channel to its "brains," for example by attaching a physical cable between the control box and aircraft. Thus, we know that we can transmit key information to the vehicle with no possibility that an eavesdropper will pick it up.

- Because it's easy for us to connect to the vehicle's computer—we have physical access to the vehicle whenever it's on the ground—it is feasible for us to change the encryption key every time we launch.

- The control session has a fairly limited duration (the endurance of the vehicle's power source—minutes or hours at most, not weeks or years). Recordings of control sessions are of no interest to an attacker—he needs to subvert a control session while it's actually in progress in order to achieve his goals.

- We have good physical control over all components of the cryptosystem, so we don't need to be overly concerned that someone could steal a piece of equipment with a valuable key in it. Any key information stolen this way is worthless, because it relates only to a past communication session.

With all this information in hand, a reasonable choice of cryptosystem for this application is a moderate-security (say, 64-bit) symmetric algorithm, optimized for speed. The complexity of the algorithm should be chosen to strike a balance between computational resources available on board the vehicle, and the computational power we believe the attacker can bring to bear during the time period of a typical communications session. (In other words, if we were designing some advanced radio-controlled solar plane that could stay aloft for weeks, we should choose a stronger key width than for a typical plane that will only fly for an hour or so without recharging). Furthermore, in order to guard against the possibility that an attacker might intercept one communications session, take it home and cryptanalyze it at leisure, we should use a different, random key every time we launch the aircraft.

5.4 Protecting One-Way Telemetry

A one-way telemetry link is an interesting reversal of the scenario described in the previous section. The difference between telemetry information and control information is that telemetry frequently remains valuable long after it's collected, which control information (generally) does not. In this case, we may be relying on the cryptosystem to provide both authentication (verifying that the telemetry we're receiving is actually coming from the source it's supposed to be coming from) and encryption (making sure that other people can't use our collected data). An example of this sort of application might be stock control using handheld wireless transmitters. You want to be sure that only authorized personnel can check stock out of inventory; you also want to avoid broadcasting the exact contents of your warehouse to everyone in the neighborhood.

Again, let's look at our requirements. Once more, we have a relatively low-powered handheld transmitter, but it's feasible that it could be a reasonably speedy 32-bit part, perhaps an ARM7 microcontroller with an LCD controller on-chip. Let's assume, however, that it is too slow to implement an asymmetric algorithm. It is probably safe to assume also that we can collect the transmitters at the end of every day and perform some physical link to them. Our aim, for the sake of argument, is to prevent the competitor across the road from intercepting our shipment orders and deducing which products we're selling briskly. (We're in a cut-throat business. If our competitor finds out that our left-handed widgets are selling quickly, he might choose to undercut our price, even if it means a net loss to him, and drive us out of the market. Or if he sees that we're using a huge quantity of some particular part, maybe he'll try to buy up stocks of that part and raise the market price to damage our operations). A small amount of data leakage is acceptable.

We can satisfy all our requirements with a system that comprises the following features:

- The transmitters use a symmetric-key algorithm with a key width that's reasonably hard to crack with commercial-grade computational power.

- Each transmitter has a serial number that can be read out using a physical connection to the unit.

- Employees are instructed to put the transmitters onto charge/reprogramming stations after every shift.

- Each unit is loaded with a new random key when it is put on the charge station. The station interrogates the unit to find out its serial number, and informs the central computer (over a secure, wired link) of the serial number and the assigned key. No mechanism is provided for the current key to be read out of the unit.

- Every transmission from the unit is encrypted with the key assigned for this specific unit for this shift. Since this is constantly changing, if our attacker happens to break a particular key, he can only recover one shift's worth of messages from one handheld unit.

- The stock-control computer is off-site. All stock add/remove requests are forwarded to the stock-control computer verbatim; that is, the local receiver hardware does not remember assigned keys, and there is no on-site information to decrypt those on-air messages.

Note that I haven't explicitly discussed the cryptosystem that protects the link between this warehouse and the central computer; I've assumed that it's strong and reliable. One good choice would be to use an asymmetric algorithm, where the random-key-generator box in the warehouse uses the central computer's public key to encrypt its reports on which keys have been assigned to which units.

5.5 Protecting Bidirectional Control/Data Streams

Many of the sorts of links you'll deal with will be fully bidirectional. For instance, you might have an application with an embedded web server that can be used to control the appliance as well as retrieving data from it. Protecting systems of this sort is an interesting topic with several solutions, depending on what your network looks like and the level of security you require versus the degree of annoyance you are willing to endure.

Probably the best way of securing your data link (short of a one-time code pad) is to use a wide-key symmetric cryptosystem. It's fast, it's secure—it works very well. The problem is that key management is difficult—if you have one single key that's used for all appliances, that key becomes a very tempting target and an appallingly risky single point of failure. On the other hand, if you have a different key for every appliance you talk to, managing all those keys becomes a big chore. Furthermore, you have to find some way of delivering those keys securely, which puts you almost back at square one, looking for a secure communications channel.

A good second best—potentially more secure, but not always feasible—is to use an asymmetric-key algorithm. At the start of the communications link, the two parties exchange public keys, and use the other person's public key to encrypt data they are sending, and their own private key to decrypt data they are receiving. This technique is, however, usually avoided due to the high computation requirements of asymmetric-key algorithms with reasonably wide keys.

One system that works around this issue quite well is to use a combination of asymmetric- and symmetric-key encryption. This system is frequently used for Internet communications protocols; in fact, I wrote the encryption system for a VPN tunneling package, using this type of methodology.

The way it works is as follows: Let us imagine two users, Alice and Bob. Alice has a private key A and a public key a. Bob has a private key B and a public key b. In real implementations, A, a, B and b are frequently random, and are sometimes generated immediately before a connection is established. To begin a communications session, Alice first sends a to Bob. This transmission doesn't need to be encrypted in any way. Bob responds by picking a random (symmetric) session key S_B. He encrypts S_B with Alice's public key a, yielding S_B' and sends back a message that contains this S_B', along with his public key b. Anyone listening to the transaction can't work out S_B because they don't know Alice's private key A and can't feasibly deduce it from a.

At this point, Alice uses A to decrypt S_B' and thereby reconstruct a local copy of S_B. She now generates a second random symmetric session key S_A. This is encrypted with Bob's public key b to yield S_A'. Alice now sends Bob another message, containing S_A'. Bob uses his secret key B to decrypt this and reconstruct a local copy of S_B. Secret session keys have now been securely exchanged; the link is almost ready to use, but should first be tested. For some unfathomable reason, some implementations I have inspected choose to perform this link test by encrypting some known, constant piece of data (for example, "Have a nice day") and sending it across the link. This is a very serious security flaw, because it gives any attacker a free head start in cracking the session keys. A much better idea is for both Alice and Bob to generate a small block of cryptographically secure random data. They make two copies of the data; one is encrypted with the other party's public key, the other is encrypted with the appropriate session key. These double packets are then exchanged. Each party uses his own private key to decrypt the asymmetrically-encrypted copy of the random data, and the appropriate session key to decrypt the other copy. If the two copies match, then the link is known good, and the test has been carried out using a method that doesn't leak any information to an eavesdropper.

For the remainder of the session, Bob uses S_B to encrypt data he is transmitting to Alice, and S_A to decrypt data he has received from Alice. Conversely, Alice uses S_A to encrypt data she is sending to Bob, and S_B to decrypt data received from Bob. This handshaking process can be repeated as often as desired, to enhance security—in the tunneling application I mentioned, for example, new session keys were generated every 15 minutes. The algorithms being used were 2 kbit RSA and triple DES for the asymmetric and symmetric modules, respectively.

The main vulnerability of the system as I've just described it is that it doesn't protect at all against someone who sits between Alice and Bob and who can prevent them from hearing each other directly. Such an entity could pretend to be Bob when he's talking to Alice, and Alice when he's talking to Bob. You could avoid this possibility by exchanging the public keys a and b over a known-to-be-trusted channel. It doesn't have to be a secure channel (eavesdroppers are okay), it just has to be guaranteeable that there is nobody in between intercepting and modifying communications. In this way, the public key itself becomes an authentication token. At the start of each session, Alice can send Bob a test message (in plaintext), along with a hash of the message that has been encrypted with her private key A. Bob can hash the message himself, decrypt Alice's hash with her public key a, and compare the two hashes; if they match, then he is certain that he's really speaking to the owner of public key a. Similar signatures should be added to the handshaking messages described above. An entity between Alice and Bob will not know their private keys and will be unable to fake these messages. Given a secure hash algorithm, he will also be unable to fake out the test message contents in such a way as to generate the correct encrypted hash.

5.6 Protecting Logged Data

Consider a project like E-2, or perhaps more accurately consider the probable specifications of a government-sponsored version of such a device. If you're sending a robot to perform surveillance duties, it's very important that the data it records should not be recoverable by a third party. This is a very interesting problem. We're not merely protecting some ephemeral data link against attack—we have to assume that the vehicle itself will fall into enemy hands. We want to ensure that they can't discover what the vehicle learned. We would also like to avoid the possibility that an enemy could capture the vehicle, overwrite its log with falsified information, and then send the vehicle back on its way to deliver fake information to us.

Note that it is not a complete solution simply to move the logging function into our monitoring station and out of the vehicle itself. If the enemy intercepts and records the data link, then captures the vehicle, they've got all the time in the world to recover the keys and decrypt their transcript of the telemetry uplink. Besides, in some applications (submarines, for instance!) it's very difficult to establish a guaranteed real-time telemetry link back to home base.

This fact immediately leans us away from symmetric-key algorithms. If we were using a symmetric-key system, we would have to have the key itself stored in the appliance, ready for an attacker to recover. There are some specialized processes (chemical security coatings for the dice; these coatings react to light or atmospheric exposure and destroy the chip contents) that can be applied to cryptographic micro-processors and ASICs to prevent key recovery, but they're very expensive and there's a risk that they could be defeated.

A better approach is to use an asymmetric algorithm, where the logging device knows a public key, which is used to encrypt all stored data. Anyone who recovers the unit, even if they tear down the hardware and reverse-engineer it fully, will not be able to recover or deduce the matching private key. The problem now becomes one of authentication. How can we be sure that the enemy hasn't captured the device, reverse-engineered it and generated a fake log using the public key that was stored in it? This is a much tougher nut to crack, and it will most likely ultimately boil down to some level of hardware security. For example, you can have the log data run through a piece of separate hardware that signs the log entries before they are stored to disk. This piece of hardware can be buried (physically) deep inside the appliance. Intrusion sensors can then be used to detect reverse-engineering and destroy the contents of the signature module. Hardware like this is often also time-sensitive—it requires all communications to be on a regular schedule, otherwise it self-destructs. This prevents an enemy from freezing the system and gaining leisure time to think about how to attack it.

It's also vital, in an application like this, to ensure that sensitive information isn't stored temporarily in unencrypted form. For instance, we might be using a digital camera to capture images into RAM; they are then compressed, encrypted and stored on a hard drive. An attacker could open the device, freeze the microprocessor (by halting the clock signal) and use a logic analyzer to read out the contents of the RAM. Protecting against these sorts of issues tends to become a matter of simply closing windows as quickly as possible. In the specific case I just mentioned, you should compress and encrypt the image immediately it is acquired, then erase the unencrypted buffer.

If you are using an operating system that implements virtual memory, you should also make absolutely certain that memory used for sensitive data does not have virtual memory behind it. Secure operating systems are designed to take these issues into account implicitly.

5.7 Where to Obtain Encryption Algorithms

Linux kernel 2.4.24 includes a comprehensive cryptographic subsystem with numerous algorithms pre-implemented and tested for you.

- MD4 (RFC1320) and MD5 (RFC1321) digest algorithms.

- SHA1 (FIPS 180-1/DFIPS 180-2) hash algorithm.

- SHA256, SHA384 and SHA512 (DFIPS 180-2) hash algorithms.

- DES (FIPS 46-2) and Triple DES EDE (FIPS 46-3). DES is a rather hoary old 56-bit symmetric-key cryptosystem, formerly considered adequate for civilian communications. Except for backwards compatibility with other products, DES should be considered uselessly obsolete—AES, below, was intended to replace it.

- Blowfish, a 32 to 448-bit symmetric-key cipher.

- Twofish, a 128/192/256-bit symmetric-key cipher.

- Serpent, an 0 to 256-bit symmetric-key cipher.

- The FIPS-197 AES algorithms, i.e., Rijndael with key sizes of 128, 192 or 256 bits.

- CAST5/CAST-128 (RFC2144) symmetric-key cipher.

Asymmetric-key cryptosystems are conspicuously absent from the above list. (This appears to be more because of patent restrictions than government regulation). You may want to visit *http://www.thefreecountry.com/sourcecode/encryption.shtml*, where ready-to-run sourcecode for many popular algorithms is available for you to download.

Warning: Many, if not all, of these algorithms are patented. You should consult local fair-use legislation before using them for any commercial or publicized purpose. Private research is *usually* covered by fair-use laws and can generally be pursued without fear of reprisal, but in some cases (DMCA again!) even private research is prohibited.

CHAPTER 6

Expecting the Unexpected

6.1 Introduction

You'll recall that in the introduction, I said that my target readership is familiar with either Linux application programming or embedded development. This chapter is mainly aimed at the former category of reader; most embedded developers should be familiar with most of the material in here.

In this chapter, I'll describe a little of the engineering behind fault detection and mitigation. More specifically, I'll talk a bit about the fault detection and failsafe mechanisms I have put in E-2. There are numerous excellent references on the more general topic, and if you read them you'll be struck by the loss of life and financial costs of the anecdotes they use to illustrate their examples. Two reports that you'll find to be most interesting reading (they are the usual starting point for discussions of software reliability) are the report on the demise of the European Space Agency's first Ariane-5 rocket, and the report on the failures of the Therac-25 radiotherapy units. A quick web search on either of those topics will lead you to the original reports.

Failures in E-2's software and firmware won't bring down any national budgets or kill anyone, but loss of the craft does represent a huge financial setback for me personally. As a result, the firmware is structured towards recovery of the vehicle after any failure. This reflects my particular design priorities. If this were a government project, it would quite possibly be designed with data security as its first priority—the hardware would be considered expendable.

6.2 Dangerous Exception Conditions and Recovering From Them

In analyzing how to protect a system against entering unknown or illegal states, you will need to create a list of things—voltages, memory variables, and so on—that can be monitored. Of only slightly lesser importance than analyzing the device's possible failure modes, however, is to decide exactly how to recover from a problem detected in one of the parameters you're monitoring. A typical analysis would identify the following:

- The parameter to be monitored. For example, this might be an analog voltage (perhaps corresponding to some real-world measurement such as temperature), the state of a variable (a counter, for instance).

- A range of values for which the parameter is considered within normal operating limits.

- A range of values for which the system behavior should be temporarily constrained in some way, and a clearly defined recovery methodology. For instance, if a battery is outside its recommended charge temperature range, the system should not enable charging. On the other hand, this is not necessarily an error; the battery may just have been brought in from a cold environment, or something of the kind. The system should allow some period of time before declaring a fatal error condition.

- A range of values for which the system should be partly or wholly shut down, and the operator (if any) informed of a serious problem.

- Analysis of which system functions can still be provided if a partial shutdown occurs. For instance, if your car's ECM detects engine sensor problems, it can switch into an emergency "limp-home" mode, where it operates with considerably reduced fuel efficiency or other undesirable behavior, but it can at least function sufficiently well to get you off the highway.

- An estimate of the time available between an excursion from normal values and a physical system problem (explosion of a battery, for instance!)

- Preferably, a means to cross-verify that the value being read corresponds to the actual system state.

■ External interlocks that can clamp related signals or effects automatically if the given parameter goes out of range.

E-2 monitors a large amount of environmental information as part of its normal mission profile. Some of this information can be used to determine if the system is in danger. Most of the danger conditions (for our limited definition of the word "danger," anyway) occur when the vessel is completely submerged. For this reason, the focus in E-2 is on bringing the vehicle to the surface, if possible. If that's not possible, the secondary emphasis is on advertising the vessel's location so that it can be recovered. There is a module dedicated entirely to energy management and vehicle recovery; it has its own independent power supply.

Here's a list of some of the things we monitor and recovery steps we take:

■ The absolute external water pressure, and the differential pressure across the hull. The vehicle has an emergency canister of carbon dioxide (connected to the interior compartment of the boat via a solenoid valve) which can be used to pressurize the hull and expel water. If the pressure differential across the hull exceeds rated limits, we add gas pressure to the boat to reduce water leaks. If the exterior pressure falls below the interior pressure, we open a second valve in the keel to release gas pressure. This prevents the vehicle from causing injuries when it's opened at the surface.

■ Internal bilge sensors. Some water in the bottom of the boat is inevitable, but if it rises above a certain threshold level, the CO_2 cylinder is fired, the keel valve is opened, and the boat is commanded to surface.

■ System battery state. The vessel has a main battery, used to power it for most of the mission, and a reserve battery that can be used for emergency maneuvers. If the control module detects that the main battery is low, it aborts whatever activity is in progress, disconnects nonessential modules (camera, SBC, and so on) from the power bus, and switches to the reserve battery. The vehicle is then commanded to surface; dive planes are brought to a mild rising angle, the rudder is straightened, and the motors are commanded to half-speed ahead.

- Internal temperature of motors and battery compartment. Rising motor temperature indicates a friction problem, and can abort the mission. Abnormal battery temperatures may affect their ability to deliver charge; again, if things get too far out of range, we abort automatically.

- If the system and reserve batteries are both low, or if no change is detected in exterior pressure during an emergency surfacing operation, a solenoid is triggered to release a small polystyrene-foam buoy, tethered to the vehicle by fishing line. It is hoped that this buoy can reach the surface and indicate the vehicle's position.

- Once an emergency recovery situation is declared, the recovery module disassociates itself from the vehicle's main power bus and begins transmitting an intermittent acoustic beacon and blinking an array of white LEDs (with a very low duty cycle). The recovery module's battery is calculated to operate it in this mode for about 72 hours, which should be long enough to find and recover the vessel.

Even the external light level has potential system survivability value, although in the current design this information is used only to determine whether the vehicle should turn on its exterior lights or not. A future version of the E-2 project will include solar cells for long-range missions (primarily, floating about in the middle of a body of water, collecting long-term data—the solar option is not intended to increase travel range significantly).

6.3 On-Chip vs. Off-Chip Watchdog Hardware

Most microcontrollers have an on-chip watchdog. This is a simple timer circuit that resets the micro if it does not receive some regular signal (referred to as a "kick"). The great thing about on-chip watchdogs is that they are free. The downside to them is that you're stuck with whatever the manufacturer thought suitable to implement, and this can leave a lot of gaps in your armor against runaway conditions. Here are a few common shortcomings of watchdog hardware in general:

1. **Some watchdogs can be manually disabled after they have been explicitly enabled.** This is a very bad design flaw. A good watchdog should be enabled by a register write or similar operation (once the system has finished power-on initialization), and it should be impossible for software to disable the wachdog

2. **Many on-chip watchdogs do not generate external signals when they fire.** In general, what this means is that a watchdog bite will usually not cause the microcontroller to drive its reset output network (if it has one) active. This can be a blessing or a curse. It's a curse if you want the watchdog bite to lead guaranteeably to a fully-reset system configuration; you have to dedicate an I/O pin to providing a "reset out" signal.

3. **Some on-chip watchdogs accept uselessly broad kick conditions.** For instance, they might regard any write to a range of ports as a valid kick. It's better to have a watchdog that requires at least two sequenced writes of specific data to different addresses; that way, you can be sure that a kick is really a kick, not just a random write through a dangling pointer.

4. **All watchdogs are useless if used inappropriately.** Too many embedded programmers think they have a safe system if the watchdog is enabled and is being kicked regularly enough to keep the system from resetting. In fact, it's necessary to do some sanity checking before you kick the watchdog. This can range from simply kicking the dog once in your main loop (this works quite well in round-robin task schedulers, if you only want to protect against infinite loop conditions) to very sophisticated techniques where you measure the time spent in various different subroutines and compare this against a nominal execution profile—too much time spent in one routine, or timeslice starvation of other routines, will cause a reset. In between these two extremes are methods that check the state of a few variables and other parameters for consistency.

5. **It takes a finite time for the system to restart after a watchdog bite.** This is a very serious limitation of practically all watchdog hardware. Any safety-critical system needs to have external interlocks to mitigate this problem.

A common external hardware watchdog technique is the "pulse maintained re-lay" (PMR). E-2 uses this technique in addition to on-chip watchdog hardware. The PMR consists of a simple circuit that expects to see an AC voltage on its input. This voltage is generated by a pulse train coming out of one of the microcontroller's I/Os. If the pulse frequency falls outside a certain range, the relay opens for a specified time period, thereby interrupting the circuit's power and, hopefully, resetting the system to a known state. This is a very idiot-proof method of protecting a circuit against unexplained lockups.

You can find some interesting reading on relays in general, and more particularly, specialized relay circuits of this type, at *http://www.ibiblio.org/obp/electricCircuits/Digi-tal/DIGI_5.html*. An excellent piece of reading on microcontroller watchdogs is Niall Murphy's "Watchdog Timers" article in Embedded Systems Programming, *http://www.embedded.com/2000/0011/0011feat4.htm*.

6.4 Good Power-On Reset Practices

At a rough guesstimate, something like 75% of hobbyist and commercial micro-controller circuits generate their power-on reset (POR) signals using a simple RC network. An example of this sort of configuration is shown in Figure 6-1 (this circuit is correct for an active-low reset signal).

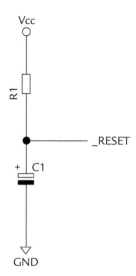

Figure 6-1: Simple POR circuit

It is assumed that the capacitor is completely discharged at the moment when the appliance is switched on. At power-up, Vcc (theoretically) rises instantly to its nominal value, so the microcontroller should be powered-up immediately. The capacitor holds the reset pin low until the current flowing through the resistor has charged it up to the input pin's logic high threshold. Thus, the length of time the reset signal is active depends on the time constant RC, the voltage Vcc, and the specified logic threshold value of the microcontroller's reset input pin.

What's wrong with this configuration? Well, the first thing to consider is that I lied shamelessly to you in the preceding paragraph. The active pulse width on the reset signal also depends on the characteristics of the input pin(s) to which the RC network is connected. If you connect more than one input pin to a single RC network, the overall behavior will deviate further and further from the calculated ideal. On the other hand, if you use a separate RC network for each section of the circuit that requires a power-on reset, you'll inevitably have different parts of the appliance coming out of reset at different times. Thus, an improvement on the circuit in Figure 6-1 would be to run the signal into a buffer (typically a NAND gate, or one or two inverters are used, depending on whatever discrete gates happen to be spare in the circuit being constructed), and to fan out the buffered reset output to whatever parts of the circuit need a reset signal. For example, the following:

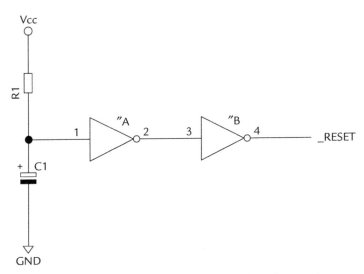

Figure 6-2: Slightly refined POR schematic

The second thing to keep in mind is that the active time isn't the only important parameter on the reset signal. All logic inputs have a maximum rise/fall-time specification, which you'll find in the device's datasheet. Recall that the V/t charge curve for a capacitor is exponential in nature; it rises very quickly from zero, but flattens off and, in theoretical terms, will never actually reach Vcc. What this means is that depending on the specific values of your resistor and capacitor, it's possible that the micro may see an abnormally slow risetime around the logic threshold voltage. This situation is exacerbated by the fact that the power rail itself exhibits less-than-ideal behavior. Some local slumping can be expected, particularly since at power-on and during brownouts, the supply rail is heavily loaded by the need to charge up all the bypass capacitors on the board. In practical terms, then, it's best to choose C to be large and R to be small, so that when the voltage crosses the critical logic threshold, the capacitor is still in the steep early regions of its charge curve—even if Vcc is actually a bit lower than its nominal value. A slightly more complete solution is to use a buffer with Schmitt trigger inputs. This will ensure that the logic level presented to the microcontroller is always a clean state.

One partial workaround for these shortcomings—and I must admit that I've been guilty of perpetrating this in commercially fielded products—is to wire the product's power switch so that, in the "off" position, it shorts out Vcc to ground. This helps the situation in normal-usage circumstances because it ensures that the POR capacitor is fully discharged very shortly after the power is turned off. The inadequacy of this workaround lies in the fact that not all potential power failures are caused by a user flipping the power switch on the device. Temporary interruptions (blackouts) or slumps (brownouts) in the mains supply voltage can, for mains-powered appliances, simulate a power-on condition without anyone ever touching the power switch. If these interruptions are short, the capacitor in our reset network won't have time to discharge fully, and it will consequently charge up over the logic threshold faster than we expect. In a worst-case scenario, a brief brownout or blackout will lower Vcc below the micro's operating threshold, but won't allow the capacitor to discharge far enough to generate a proper reset pulse when power is restored. Pretty much anything could be happening to the micro in this scenario; it could be running normally (albeit with no I/Os because of a depressed I/O ring voltage), it could be frozen, it might be executing out of unimplemented ROM space, or it might have reached some undefined internal state where it can't execute any code at all until it receives an external reset signal.

A carefully-structured POR circuit is, therefore, integral with a brown-out detector. It should assert the reset signal when power is applied to the system. Ideally, reset should be asserted before the microcontroller is powered up, and the POR circuit should hold the signal active until the power rail is at a nominal value. Furthermore, our mythical POR circuit should detect brownout conditions on the power rail, and should supply a clean, known-width reset pulse if such a condition occurs. Fortunately, we don't need to design a chip to do this. Maxim, for example (*http://www. maxim-ic.com/*), sells several appropriate devices, and they're very cheap. E-2 uses these integrated power-on-reset generators/brownout detectors extensively.

6.5 A Few Additional Considerations for Battery-Powered Applications

Battery-powered appliances, in the main, need to exercise particular care over how they detect and handle hardware exception conditions. Special rules apply to error recovery, because it's possible that you might not have enough life left to get all the way through a recovery algorithm. When exceptions occur in a battery-powered device, your first priority should be to get the system into a state where it will be safe if the microcontroller goes completely offline. Systems operating off battery power constantly live under the Sword of Damocles; they scurry nervously from one safe state to the next, with as little time as possible spent in between. In E-2's case, the most worrying time for us is when the keel valve is open for any reason; it's latched, to save power, and we might not have enough energy to close it again.

Another consideration which affects most devices that use rechargeable batteries, is that these batteries will typically be damaged if they are discharged below a certain cell voltage. It is normal, in such circuits, to set up a low-battery warning that gives the system a known grace period to shut down, and then for the microcontroller or an external power supply circuit to shut the system down explicitly when a critical battery level is reached. Not only does this protect your batteries against over-discharge, it also allows the system to shut down important systems gently and elegantly. Note that one potential problem with this system occurs if the user powers off the device, then switches it back on once the batteries have had time to accumulate a surface charge. These batteries are already further down their discharge curve than they appear (from simple voltage measurements). The unit may not have as much time as it thinks between "low battery" and actual death. The best way to mitigate this problem is by including a gas gauge function in the battery itself, so that the unit cannot be powered up again until the battery is swapped out or charged. Cellphones and laptops frequently implement this sort of system.

And finally, while we're talking about battery-powered appliances, you should be particularly careful about implementing charge controller features (for rechargeable batteries) entirely in software. If you do use a microcontroller to perform charge control, the code must be rigorously designed and carefully debugged—and you should have an external hardware interlock as well (thermal fuses to protect against over-temperature, regular fuses to protect against overcurrent, and so on).

# CHAPTER 7

Contents of the Enclosed CD-ROM

Item	Path	Description
AVR Studio 4.08	/utils/AVR Studio 4.08/	The Atmel AVR Studio development environment for Windows.
Busybox	/linux/busybox-0.60.5.tar.gz	The Busybox utility package for Linux.
EAGLE (Linux)	/utils/eagle-4.11e.tgz	EAGLE PCB CAD package for Linux.
EAGLE (Windows)	/utils/eagle-4.11e.exe	EAGLE PCB CAD package for Windows.
Linux kernel	/linux/linux-2.4.24.tar.gz	Sourcecode for Linux kernel 2.4.24.
Linux kernel configuration	/linux/geode-config	Configuration file for Linux kernel 2.4.24 on Advantech PCM-5820 or compatible Geode-based SBC.
LIRC	/linux/lirc-0.6.6.tar.gz	Sourcecode for LIRC infra-red driver.
Sample programs for Linux	/linux/sample-programs.tar.gz	Contains entire source tree for all sample Linux programs mentioned in this text.
Sample hardware project schematics	/projects	Schematics and firmware for circuits described in this book. These are all in EAGLE format.
EAGLE libraries for hardware projects	/projects/libraries	These library files contain parts not found in the standard EAGLE libraries.
Sample root filesystem for CompactFlash or CD-ROM boot	/card-root.tar.gz /cdrom-root.tar.gz	A complete root filesystem as created by the steps described in Sections 4-4 and 4-5.
SVGAlib	/linux/svgalib-1.4.3.tar.gz /linux/svgalib-1.4.3-patched.tar.gz	The SVGAlib graphics library sourcecode. The -patched archive has been patched to build correctly with gcc 3.x.

Index

Symbols
0xFF, 148
1024-bit RSA, 213
3.5" biscuit, 22
5.25" biscuit, 22
8051 architecture, 14, 15

A
acceleration sensor, 80
accelerometer schematic, 84
acquiring image data from cameras, 189
active pulse width, 233
active time, 234
Advantech's BIOS 1.23, 63
Advantech PCM-5820 Single-Board Computer, 27
ADXL202, 82
  ADXL202 output signals, 83
  ADXL202 variants, 82
ADXL202JQC, 81
algorithms, 214, 215
  asymmetric-key algorithm, 220
ANSI C compiler, 17
ARM-Linux, 21
ARM7-cored microcontrollers, 21
assemblers, 13
asymmetric-key
  algorithms, 215
  cryptosystems, 224
  systems, 216
  encryption, 220
asymmetric algorithm, 218, 223
asymmetric and symmetric modules, 221
asynchronous serial port, 18
Atmel AVR, 14, 15, 17
  ATtiny26L, 17, 20, 48
attitude sensor, 80
AVR chip, 17
AVR fuse settings, 19
AVR Studio, 19, 21
AVR Studio 4.08, 237
AwardMod, 206
Award BIOS, 136
Award Modular BIOS, 202
AWFLASH, 204

B
Battery-powered appliances, 235, 236
BIOS, 136, 141, 202, 203, 204, 206
BIOS customization utilities, 208
blackout, 234
bootable CD-ROMs, 141
bootable CompactFlash image, 120
bootable disk, 141
bootable filesystem, 116
bootable system restore, 117
boot image, 141
brown-out detector, 234, 235
brown-out conditions, 235
build and test your embedded kernel, 120
Busybox, 237
  program, 130

C
capacitor, 233
CD-ROMs, 137
central system controller, 117
clock input (SCK), 49
closed-source encryption products, 213
closed-source product, 212
CMOS, 144
  image sensors, 10
color cameras, 189
CompactFlash card, 23, 116, 126, 127, 123
  startup, 120
  storage media, 10
compilers, 13
configuring the development system, 117
control-critical data transfers, 11
core clock source, 18
CPU modules, 16
creating a bootable linux system-restore CD-ROM disc, 136
creating a root filesystem for an embedded
    system, 128
creating a custom kernel, 117
cryptosystem, 211, 212, 216, 218
customizing the BIOS, 201
custom BIOS, 201
custom CMOS default settings, 201

ELSEVIER SCIENCE CD-ROM LICENSE AGREEMENT

PLEASE READ THE FOLLOWING AGREEMENT CAREFULLY BEFORE USING THIS CD-ROM PRODUCT. THIS CD-ROM PRODUCT IS LICENSED UNDER THE TERMS CONTAINED IN THIS CD-ROM LICENSE AGREEMENT ("Agreement"). BY USING THIS CD-ROM PRODUCT, YOU, AN INDIVIDUAL OR ENTITY INCLUDING EMPLOYEES, AGENTS AND REPRESENTATIVES ("You" or "Your"), ACKNOWLEDGE THAT YOU HAVE READ THIS AGREEMENT, THAT YOU UNDERSTAND IT, AND THAT YOU AGREE TO BE BOUND BY THE TERMS AND CONDITIONS OF THIS AGREEMENT. ELSEVIER SCIENCE INC. ("Elsevier Science") EXPRESSLY DOES NOT AGREE TO LICENSE THIS CD-ROM PRODUCT TO YOU UNLESS YOU ASSENT TO THIS AGREEMENT. IF YOU DO NOT AGREE WITH ANY OF THE FOLLOWING TERMS, YOU MAY, WITHIN THIRTY (30) DAYS AFTER YOUR RECEIPT OF THIS CD-ROM PRODUCT RETURN THE UNUSED CD-ROM PRODUCT AND ALL ACCOMPANYING DOCUMENTATION TO ELSEVIER SCIENCE FOR A FULL REFUND.

DEFINITIONS

As used in this Agreement, these terms shall have the following meanings:

"Proprietary Material" means the valuable and proprietary information content of this CD-ROM Product including all indexes and graphic materials and software used to access, index, search and retrieve the information content from this CD-ROM Product developed or licensed by Elsevier Science and/or its affiliates, suppliers and licensors.

"CD-ROM Product" means the copy of the Proprietary Material and any other material delivered on CD-ROM and any other human-readable or machine-readable materials enclosed with this Agreement, including without limitation documentation relating to the same.

OWNERSHIP

This CD-ROM Product has been supplied by and is proprietary to Elsevier Science and/or its affiliates, suppliers and licensors. The copyright in the CD-ROM Product belongs to Elsevier Science and/or its affiliates, suppliers and licensors and is protected by the national and state copyright, trademark, trade secret and other intellectual property laws of the United States and international treaty provisions, including without limitation the Universal Copyright Convention and the Berne Copyright Convention. You have no ownership rights in this CD-ROM Product. Except as expressly set forth herein, no part of this CD-ROM Product, including without limitation the Proprietary Material, may be modified, copied or distributed in hardcopy or machine-readable form without prior written consent from Elsevier Science. All rights not expressly granted to You herein are expressly reserved. Any other use of this CD-ROM Product by any person or entity is strictly prohibited and a violation of this Agreement.

SCOPE OF RIGHTS LICENSED (PERMITTED USES)

Elsevier Science is granting to You a limited, non-exclusive, non-transferable license to use this CD-ROM Product in accordance with the terms of this Agreement. You may use or provide access to this CD-ROM Product on a single computer or terminal physically located at Your premises and in a secure network or move this CD-ROM Product to and use it on another single computer or terminal at the same location for personal use only, but under no circumstances may You use or provide access to any part or parts of this CD-ROM Product on more than one computer or terminal simultaneously.

You shall not (a) copy, download, or otherwise reproduce the CD-ROM Product in any medium, including, without limitation, online transmissions, local area networks, wide area networks, intranets, extranets and the Internet, or in any way, in whole or in part, except that You may print or download limited portions of the Proprietary Material that are the results of discrete searches; (b) alter, modify, or adapt the CD-ROM Product, including but not limited to decompiling, disassembling, reverse engineering, or creating derivative works, without the prior written approval of Elsevier Science; (c) sell, license or otherwise distribute to third parties the CD-ROM Product or any part or parts thereof; or (d) alter, remove, obscure or obstruct the display of any copyright, trademark or other proprietary notice on or in the CD-ROM Product or on any printout or download of portions of the Proprietary Materials.

RESTRICTIONS ON TRANSFER

This License is personal to You, and neither Your rights hereunder nor the tangible embodiments of this CD-ROM Product, including without limitation the Proprietary Material, may be sold, assigned, transferred or sub-licensed to any other person, including without limitation by operation of law, without the prior written consent of Elsevier Science. Any purported sale, assignment, transfer or sublicense without the prior written consent of Elsevier Science will be void and will automatically terminate the License granted hereunder.

TERM

This Agreement will remain in effect until terminated pursuant to the terms of this Agreement. You may terminate this Agreement at any time by removing from Your system and destroying the CD-ROM Product. Unauthorized copying of the CD-ROM Product, including without limitation, the Proprietary Material and documentation, or otherwise failing to comply with the terms and conditions of this Agreement shall result in automatic termination of this license and will make available to Elsevier Science legal remedies. Upon termination of this Agreement, the license granted herein will terminate and You must immediately destroy the CD-ROM Product and accompanying documentation. All provisions relating to proprietary rights shall survive termination of this Agreement.

LIMITED WARRANTY AND LIMITATION OF LIABILITY

NEITHER ELSEVIER SCIENCE NOR ITS LICENSORS REPRESENT OR WARRANT THAT THE INFORMATION CONTAINED IN THE PROPRIETARY MATERIALS IS COMPLETE OR FREE FROM ERROR, AND NEITHER ASSUMES, AND BOTH EXPRESSLY DISCLAIM, ANY LIABILITY TO ANY PERSON FOR ANY LOSS OR DAMAGE CAUSED BY ERRORS OR OMISSIONS IN THE PROPRIETARY MATERIAL, WHETHER SUCH ERRORS OR OMISSIONS RESULT FROM NEGLIGENCE, ACCIDENT, OR ANY OTHER CAUSE. IN ADDITION, NEITHER ELSEVIER SCIENCE NOR ITS LICENSORS MAKE ANY REPRESENTATIONS OR WARRANTIES, EITHER EXPRESS OR IMPLIED, REGARDING THE PERFORMANCE OF YOUR NETWORK OR COMPUTER SYSTEM WHEN USED IN CONJUNCTION WITH THE CD-ROM PRODUCT.

If this CD-ROM Product is defective, Elsevier Science will replace it at no charge if the defective CD-ROM Product is returned to Elsevier Science within sixty (60) days (or the greatest period allowable by applicable law) from the date of shipment.

Elsevier Science warrants that the software embodied in this CD-ROM Product will perform in substantial compliance with the documentation supplied in this CD-ROM Product. If You report significant defect in performance in writing to Elsevier Science, and Elsevier Science is not able to correct same within sixty (60) days after its receipt of Your notification, You may return this CD-ROM Product, including all copies and documentation, to Elsevier Science and Elsevier Science will refund Your money.

YOU UNDERSTAND THAT, EXCEPT FOR THE 60-DAY LIMITED WARRANTY RECITED ABOVE, ELSEVIER SCIENCE, ITS AFFILIATES, LICENSORS, SUPPLIERS AND AGENTS, MAKE NO WARRANTIES, EXPRESSED OR IMPLIED, WITH RESPECT TO THE CD-ROM PRODUCT, INCLUDING, WITHOUT LIMITATION THE PROPRIETARY MATERIAL, AN SPECIFICALLY DISCLAIM ANY WARRANTY OF MERCHANTABILITY OR FITNESS FOR A PARTICULAR PURPOSE.

If the information provided on this CD-ROM contains medical or health sciences information, it is intended for professional use within the medical field. Information about medical treatment or drug dosages is intended strictly for professional use, and because of rapid advances in the medical sciences, independent verification f diagnosis and drug dosages should be made.

IN NO EVENT WILL ELSEVIER SCIENCE, ITS AFFILIATES, LICENSORS, SUPPLIERS OR AGENTS, BE LIABLE TO YOU FOR ANY DAMAGES, INCLUDING, WITHOUT LIMITATION, ANY LOST PROFITS, LOST SAVINGS OR OTHER INCIDENTAL OR CONSEQUENTIAL DAMAGES, ARISING OUT OF YOUR USE OR INABILITY TO USE THE CD-ROM PRODUCT REGARDLESS OF WHETHER SUCH DAMAGES ARE FORESEEABLE OR WHETHER SUCH DAMAGES ARE DEEMED TO RESULT FROM THE FAILURE OR INADEQUACY OF ANY EXCLUSIVE OR OTHER REMEDY.

U.S. GOVERNMENT RESTRICTED RIGHTS

The CD-ROM Product and documentation are provided with restricted rights. Use, duplication or disclosure by the U.S. Government is subject to restrictions as set forth in subparagraphs (a) through (d) of the Commercial Computer Restricted Rights clause at FAR 52.22719 or in subparagraph (c)(1)(ii) of the Rights in Technical Data and Computer Software clause at DFARS 252.2277013, or at 252.2117015, as applicable. Contractor/Manufacturer is Elsevier Science Inc., 655 Avenue of the Americas, New York, NY 10010-5107 USA.

GOVERNING LAW

This Agreement shall be governed by the laws of the State of New York, USA. In any dispute arising out of this Agreement, you and Elsevier Science each consent to the exclusive personal jurisdiction and venue in the state and federal courts within New York County, New York, USA.